Why Don't Cats Like to Swim?

ALSO BY DAVID FELDMAN

Do Elephants Jump?

How Do Astronauts Scratch an Itch?

What Are Hyenas Laughing at, Anyway?

How to Win at Just About Anything

How Does Aspirin Find a Headache?

Are Lobsters Ambidextrous?
   (formerly published as *When Did Wild Poodles Roam the
   Earth?*)

Do Penguins Have Knees?

Why Do Dogs Have Wet Noses?

When Do Fish Sleep?

Who Put the Butter in Butterfly?

Why Do Clocks Run Clockwise?

Why Don't Cats Like to Swim?

An Imponderables® Book

DAVID FELDMAN

Illustrations by Kassie Schwan

Perennial Currents

An Imprint of HarperCollinsPublishers

First HarperPerennial edition published 1987.

First Perennial Currents edition published 2004.

Designed by Patrice Fodero

Library of Congress Cataloging-in-Publication Data
Feldman, David.
    [Imponderables]
    Why don't cats like to swim? / David Feldman ; illustrations
by Kassie Schwan.
        p. cm.
    "An Imponderables book."
    Originally published as: Imponderables. New York : Perennial, 1987.
    ISBN 0-06-075148-7 (pbk.)
    1. Questions and answers. I. Title.

AG195.F48 2004
031.02—dc22
                                                        2004044789

08 FOLIO/QM 20 19 18 17 16 15 14 13

For my parents,

Ray and Fred Feldman—the best

Foreword

You are on a diet. Forsaking your beloved bacon and eggs, you settle for Special K and skim milk. Your kid brother chuckles at your martyrdom. "Look," he points out, reading the nutritional labels of the boxes. "Your Special K has no fewer calories per serving than my Count Chocula." Your snotty brother is right. *How can heavily sugared cereals such as Count Chocula or Cap'n Crunch be no more fattening than "adult" cereals like Total or Special K? Why do Kellogg's Corn Flakes and Kellogg's Sugar Frosted Flakes both have 110 calories per ounce?*

You are so disgusted with the Special K, you sneak away to the local coffee shop for a *real* breakfast. The conservative man in the booth across from you orders soup. The waiter brings him a tray of crackers, and the man takes out a saltine package. But after several tries, he still can't open it. Furtively, he looks around the room to see if anyone is watching him. You, having seen Humphrey Bogart movies, evade his glance. You then watch him ferociously tear open the saltine package with his teeth, and you wonder: *If we can put a man on the moon, why can't they make packages you can open up with your hands instead of your teeth?*

After polishing off the danish that concludes your greasy

breakfast, you head out for work. You are zipping along the turnpike when you suddenly find yourself stuck in bumper-to-bumper traffic. You barely move for a half hour. Suddenly, traffic clears. There is no evidence of a traffic accident or a stalled car. You pass no clogged on-ramps or off-ramps or any other explanation for the tie-up. *What caused the traffic jam? What caused the traffic to clear?*

There are plenty of books that profess to tell you the meaning of life, but where can you go to find the answers to these little mysteries of modern life, the really important stuff?

There is only one place, and you've found it—*Imponderables*.

Imponderables are questions that cannot be answered by numbers or measurements or standard reference books. They are the kinds of questions that haunt you for hours . . . until you forget about them before you ever find their solutions. Repressing these knotty Imponderables might be a temporary solution, but you will recall them at the oddest moments, and they can torment you for the rest of your life.

We're here to help. In fact, we would love to know about any Imponderables that might have bedeviled you. On the last page of this book, you can find out how you can get *your* mysteries of life answered in the next edition of *Imponderables*. But for now, settle back and have fun with this batch.

Acknowledgments

Imponderables was so much fun to do that at times I forgot that writing a book is supposed to be a traumatic experience. For this I have to thank my collaborators, my friends and family, and the many people who served as sources for this book.

Most of all, I owe thanks to my agent, Jim Trupin, who proved that it is possible, contrary to popular belief, to be a terrific agent and a mensch simultaneously. I owe Rick Kott almost as much for sending me to Jim. Nick Bakalar's enthusiasm made this book possible. My editor, Eunice Riedel, understood what *Imponderables* was about from the onset and sensitively helped shape the book to meet that concept. Kas Schwan was more than my alter ego. She didn't just illustrate the Imponderables, she enhanced them.

Friends and family did a lot more than offer moral support, although they supplied plenty of that. They were my guinea pigs, testing ideas for Imponderables. If more than one in ten knew the correct answer to a proposed Imponderable, I threw it out. If more than one in three thought a proposed Imponderable idea was wretched, I threw it out. And most important, friends provided some of the best ideas for Imponderables. A special tribute to Susan Sherman, who came up with several terrific Imponderables when the book was just starting to bubble (fester?) in my mind. Heartfelt thanks to all of my friends who helped in so many ways: Mike Barson; Eric Berg; Jean Behrend; Brenda Berkman; Leon Bernhardt; Josephine Bishop; Sharon Bishop; Jon Blees; the whole Popular Culture department at Bowling Green State University; Pat Browne; Ray Browne; Linda Diamond; Fred Feldman; Gilda Feldman; Phil Feldman; Ray Feldman; Seth Freeman; Elizabeth Frenchman; Michele Gallery; Chris Geist; Jean Geist; Bea Gordon; Ken Gordon; Murray Gordon; Sheila Hennes; Uday Ivatury; Jo Ann Manera; Mike Marsden; Jeff McQuain; Jack Nachbar; Tom O'Brien; Pat O'Conner; Merrill Perlman; Larry Prussin; Dan Richland; Brian Rose; Leslie Rugg; Tom Rugg; Ellen Sargent; Karen Stoddard; Kat Stranger; Ed Swanson; Lorraine Vachon; Dennis Whelan; Heide Whelan; and Jon White.

There will be more thank yous in the succeeding paragraphs than in a Stevie Wonder Grammy acceptance speech, but so be it. If ever there was a book that relied on the kindness of strangers for its source material, this is it. Vice-presidents of major corporations must have better things to do than answer phone queries about why women open their mouths while applying mascara, but a surprising number of such people, with awfully responsible sounding titles, supplied answers for this book with openness, grace, and humor.

Although many other people were kind enough to supply information, the following sources furnished material that led directly to Imponderables published in this book: Guy Abruzzo; American Petroleum Institute; American Society for Testing and Materials; Herb Ames; Jim Andes, Cannon Mills; The Audubon Society; Rajat Basu; Eric Berg; Ginny Blair, Popcorn Institute; Judy Blumberg; William M. Borchard; Kyle Brenner, Baskin-Robbins; Rebecca Briggs, People Express; California Olive Industry; Helen Castle, Kellogg's; Champagne News and Information Bureau; Edith Chan; Molly A. Chillinsky, Coin Laundry Association; Tom Collins, Mobil Corporation; Bill

Cooper, Department of Transportation; Jack Cooper, Domino Sugar; Jud Crane.

Dairylea Cooperative; Charles Dale, Federal Highway Administration; Dr. Mike D'Asaro; Mary DeBourbon; Robert Deitsch; Claire Dillie; Prof. Robert J. Dinkin; Phil Dunne, New York Telephone; Eastman Kodak Co.; Roger Ebert, *Chicago Sun-Times;* Tom Elefante, Loew's Theaters; Charles Fay; Steve Feinberg; Dr. Fred Feldman; Frank Finnegan, Dellwood Foods; Thomas J. Flaherty, National Live Stock and Meat Board; Dr. Tom Flashman; Florsheim Shoes; Forster Manufacturing; Bob Goldberg, White Castle; Dr. Robert Goldberg, Center for Packaging Education; Laurie Gourley, Burger King; Bonny Graham, Wrigley Co.; Michael Guerin, City of Pasadena; Dr. Irwin Harris; Jim Hodges, American Meat Institute; Bruce Houston; Michael Howard; Mary E. Hox, Life Savers, Inc.; Dr. Gerald Imber; International Airline Passengers Association; James H. Jensen, General Electric.

W. Drew Kastner, NBC-TV; Diane Kemmelman; Walter Koob, United States Postal Service; Harry Korad, Society of Soft Drink Technologists; V. Allan Krejci, Hormel & Co.; Gene Krop, United Airlines; Monroe Lanzet, Max Factor; all the terrific people at Lathem Time Recorder Co. (Bill and Carol Lathem, Marla Paradise, and especially John Evans, who drew the technical illustration provided in the chapter on synchronized clocks); Lever Brothers; Ron Levington; Jim Lew; Denny Lynch, Wendy's; Steve Mayer, Walter Reade; Erin L. McCallon, Morrison, Inc.; Jeff McQuain; Dean Meadors, Mary Kay Cosmetics; Linda Meilan; Karen Montalto, Johnson & Johnson Products; Bob Montgomery, Nabisco Brands; Rita Morgan, Perdue Farms.

Howard Nash, Cunningham and Walsh; National Shoe Retailers Association; National Weather Service; Neutrogena; New York Police Department; Nissan Motor Corporation, USA; Tom O'Brien; Dr. George F. Odland; Al O'Leary, Department of Sanitation, New York; Otis Elevator; Tony Pappas, New York Telephone; Person & Covey; Christine Pines, Best Foods;

Wilbur Reese, Los Angeles County School System; Peggy Rogers, Best Foods; William Rusch, Holly Farms; Jim Ryan; John Ryan; Carole Shulman, Professional Skaters Guild; Simplex Time; Bill Smith, Federal Highway Administration; Pat M. Snyder, Arby's; Soap and Detergent Association; Jim Stacy, American Medical Association; Dan Stern; Bob Stewart, Bob Stewart Productions; Phyllis Straughn, General Mills.

George Thomas, Mutual Radio Network; Steven Touchband, Faichney USA; Alex Trebek; Audrey Trumbold, National Toothpick Holder's Collectors' Society; Unocal; Jane Venters, Kraft, Inc.; James Warren, Maine Sardine Council; Charles A. Winans, National Association of Concessionaires; Christen Wyatt, California Pistachio Commission; Jane Yates, Johnson & Johnson.

For every person acknowledged here, there was another person who asked that his or her statements be taken off the record. A few simply didn't want publicity, but they were outnumbered by those fearing possible reprisals by the companies for which they work. I never viewed *Imponderables* as a muckraking enterprise, but it didn't take long to realize that there is more than a little fear in corporate America if some of these topics were seen as threatening. It encourages me nevertheless that so many sources were willing to talk candidly to a writer calling out of the blue, even when they could extract no credit or glory for their efforts. This book couldn't have been written without them.

Contents

What is the difference between "partly cloudy" and "partly sunny" in a weather report?

The expression *partly sunny* was brought to you by the same folks who brought you *comfort station* and *sanitary engineer*. As a technical meteorological term, *partly sunny* doesn't exist. So while you might assume that a partly sunny sky should be clearer than a partly cloudy one, the two terms signify the same condition. You have merely encountered a weathercaster who prefers to see the glass as half full rather than half empty.

Actually, most of the meteorological terms that seem vague and arbitrary have precise meanings. The degree of cloudiness is measured by the National Weather Service and described according to the following scales:

Percentage of Cloud Cover	Term
0–30	clear
31–70	partly cloudy
71–99	cloudy
100	overcast

Where does "fair" weather fit into this spectrum? Fair weather generally refers to any day with less than a 50 percent cloud cover (thus even some "partly cloudy" days could also be "fair"). But even a cloudy day can be termed fair if the cover consists largely of transparent clouds. On days when a profusion of thin cirrus clouds hangs high in the sky but does not block the sun, it is more descriptive to call it a fair day than a partly cloudy one, since one thick cloud formation can screen more sunshine than many willowy cirrus formations.

You might also have heard the aviation descriptions of cloud cover used in weather forecasts. Here's what they mean:

Percentage of Cloud Cover	Term
0–9	clear
10–50	scattered clouds
51–89	broken sky
90–99	cloudy
100	overcast

Not many people know what the weather service means when it forecasts that there is a "chance" of rain. Precipitation probabilities expressed in vague adjectives also have precise meaning:

Chance of Precipitation	National Weather Service Term
0–20%	no mention of precipitation is made
21–50%	"chance" of precipitation
51–79%	precipitation "likely"
80–100%	will not hedge with adjective: "snow," "rain," etc.

How does the National Weather Service determine the daily cloud cover in the space age? Do they send up weather balloons? Satellites? Not quite. They send a meteorologist to the roof of a building in a relatively isolated area (airports are usually used in big cities) and have him or her look up at the sky and make a well-informed but very human guess.

When an elevator is illegally overloaded with passengers, who is criminally responsible?

When you are bored at breakfast, you read cereal boxes. When you are bored on an elevator, you read the elevator-inspection certificate, which in most localities is posted inside the elevator and includes not only emergency procedures, but specified weight and passenger capacities.

In most cities, it is a crime or a civil violation to overload an elevator. We've always wondered how these rules are enforced. Do the police conduct spot "weight traps," corraling unsuspecting hordes and putting them on cattle scales? Do those electric eyes on so-called security elevators actually do head counts, electronically signaling Interpol when there is one too many passengers in an elevator?

And what if the police do nab 11 people and 1600 pounds in an elevator designed for 10 people and 1500 pounds? Who is legally responsible? The last person to enter the elevator? The other 10 people, for allowing the illegal eleventh? And if there are only 10 people on the elevator, are you responsible for knowing the weight of your fellow passengers?

We talked to every branch of law enforcement in the elevator capital of the world, New York City, and we at *Imponderables* are pleased to inform you: Relax. No one could dig up a case, ever, where passengers were prosecuted for overloading an elevator, although such a rule is on the books. Even elevator inspectors we spoke to indicated that they wouldn't report freight elevators being overloaded and that they would be lucky if they didn't get cursed at for politely suggesting that maybe it would be a good idea not to try to cram ten refrigerators into a small elevator.

Little things like muggings and murders aren't the only reasons for law enforcement's laissez-faire attitude toward incipient elevator crime—overloading an elevator isn't particularly dangerous. Excess weight is not a common cause of elevator accidents. Most electronic elevators will simply not move if overweighted; others will not even close their doors.

The formula used for designating elevator capacities, developed by the federal government, is a bit on the arbitrary side. It isn't real complicated. Once the square footage and the technical specifications of the elevator are determined, a weight capacity is issued. Then that weight capacity is divided by 150 pounds to determine what number to list as the maximum passenger capacity. Obviously, this weight standard would not work for a convention of Overeaters Anonymous, and an elevator certainly can't sense whether there are 8 adults or 15 little kids in an elevator. The weight capacities are meant to be guidelines, although even the usefulness of guidelines is questionable when it is difficult to pack enough full-size adults in an elevator to exceed the stated limits and when it doesn't matter much, from a safety standpoint, whether the elevator is overloaded.

If there were a horrendous elevator accident, civil, not criminal, action is likely to occur, but the elevator passengers are more likely to be the plaintiffs than the defendants in such a proceeding. Most likely, injured passengers would sue the building that houses the elevator (for allowing its elevators to become overcrowded) and the manufacturer for building a defective elevator.

You are driving in bumper-to-bumper traffic on the highway. You have barely moved in the last half hour. Then, suddenly, traffic clears. There has been no traffic accident. You pass no clogged on-ramps or off-ramps or any other explanation for the tie-up. What caused the traffic jam? What caused the traffic to clear?

Chances are, you have been a victim of what traffic-flow specialists call the shock-wave effect. Highway drivers operate at peak efficiency around the 35 M.P.H. mark and are capable of performing satisfactorily at higher speeds. When traffic volume on a highway nears its designated optimum capacity, some stragglers begin driving under 35 M.P.H. and a traffic jam is born.

Otherwise speedy drivers react irrationally to a slowing down on a highway. Slower speeds, theoretically, should increase control and maneuverability, but drivers grow fearful as their pace declines. The shock-wave effect occurs because drivers look for the reason they had to slow down in the first place: they overreact to any stimuli, particularly the brake lights of cars ahead of them. A few slow drivers, at 25 M.P.H., can set off a shock-wave effect for miles behind them and, if the traffic volume is high, can create bumper-to-bumper traffic without any ostensible reason.

Why do traffic jams caused by the shock-wave effect suddenly disappear? Usually, traffic clears because there is a smaller ratio of traffic volume to capacity ahead—enough breathing room to prompt even slow-poke victims of the shock-wave effect to risk peeling away at 35 M.P.H. or more.

Why are some pistachios dyed red?

As if Iranians didn't have enough public relations problems, they have something else to answer for: They're responsible for that red gook you get on your hands after eating pistachio nuts.

Pistachios originated in the Middle East, growing wild in the deserts in ancient times. Pistachios were considered a rare delicacy and were so expensive that they were consumed mostly by royalty (the queen of Sheba was a pistachio partisan) and exported to Europe. Pistachios were grown in Greece and Sicily during the Roman Empire and remained popular in the Mediterranean, but didn't really catch on in the United States until the first great wave of immigrants from southern Europe in the 1880s.

Pistachios didn't become a mass-marketed item until the 1930s, when they were placed in vending machines, the type that is now used mostly to sell gumballs and assorted teen idol paraphernalia in front of markets and Woolworth's. The main reason pistachios were dyed bright red was to make pistachios stand out from the relatively pallid cashews and peanuts that were their main vending-machine competition.

It worked. Pistachios were a hit, even at their premium price—especially with kids, who were attracted by their shiny shells. So the choice of red was fortuitous. But why were pistachios dyed in the first place? In order to answer this question, we will have to tell you a little more than you probably want to know about the cultivation and harvesting of pistachios.

The natural color of a pistachio shell is ivory. But when pistachios are ripe and ready to be harvested (usually in September and October), the shell is encased by a thin rose-colored hull. When it ripens, the shell of the pistachio splits naturally, thus enabling the purchaser to open the shell with the fingernails rather than the bicuspids. While the pistachios are still on the tree, the rose hull protects not only the nutmeat, but the opening shell of the pistachio.

Iranian pistachios are harvested the same way now as they were a century ago. Workers knock the nuts off pistachio trees with poles. The nuts are picked off the ground by hand and thrown into burlap bags. Often, the nuts sit in these bags for weeks. The protective hull of the pistachio is removed by rubbing the nuts against rough surfaces, usually stones.

All of this manual contact with the vulnerable nut results in staining the shell. And though the stain doesn't affect the taste of the nutmeat, the resultant shell has the aesthetic appeal of a pale linen tablecloth with sweat stains. The red vegetable dye, first introduced by American importers in the 1930s, was a marketing tactic to draw attention to pistachios in vending machines and to allay consumers' fears about the blotched shells.

American farmers realized that pistachios were a potentially valuable crop, but there were several stumbling blocks to cultivation in this country. Pistachio trees are unusually sensitive to inclement weather, and it takes from seven to ten years for planting until the first yield begins. Many species of pistachios had shells that didn't seem to want to open.

Pistachios were a natural for California, which had the requisite warmth and dryness that pistachios demand. Eventually, California farmers settled on the Kerman tree—a strain of pistachios that had large nutmeats and split open naturally more often than others they researched. Pistachios were first planted commercially in the late 1960s and were first marketed in 1976.

The American farmers developed a technology not only to maximize efficiency, but to eliminate the need to dye pistachios. Pistachios are shaken off the tree by machines and never hit the ground. The nuts are immediately loaded into containers and processed: cleaned of leaves and twigs, hulled, washed and dried. The hulling and drying is accomplished so quickly that the shells have no time to stain, and they can be marketed in their natural ivory color.

It is difficult to see why anyone would want to buy the red-dyed pistachios when the naturals are available (it's a little like an M&M clone competing by boasting that they "melt in your hand, not in your mouth"), but old habits die hard. Although consumer preference is the only reason to do so, 40–50 percent of California pistachios are dyed red. The California Pistachio Commission reports that it expected consumers to switch to natural pistachios quickly but have found that East Coast customers particularly resist the change. New purchasers of pistachios, however, invariably choose naturals.

Red-dyed pistachios eventually may become a fondly remembered figment of our nostalgic past (although not too soon, since most imported pistachios are still dyed red), but it is unlikely that too many tears will be shed over the passing of the white pistachio, which can still be seen occasionally. White pistachios, after they were roasted, were coated with a mixture of salt and cornstarch that not only managed to come off on one's hands but also to mask the subtle, buttery taste of the pistachio. Even *Imponderables* can't answer why anyone would want to dye a pistachio that awful color.

Why is the color blue associated with baby boys? Why is the color pink associated with baby girls?

The association of colors with babies undoubtedly started as an attempt to identify the gender of that one group of humans to whom the cliché "they all look alike" often applies.

But why blue for boys? In ancient times, it was believed that evil spirits lingered over nurseries and that certain colors possessed the capability to combat evil. Blue was considered the most powerful color, possibly because of its association with the sky and, thus, heavenly spirits. Since boys were then considered the most valuable natural resource to parents, blue clothing was a cheap form of insurance.

Evil spirits apparently couldn't bother with pestering baby girls. Not only were girls not dressed in blue, but they had no color to call their own until centuries later. Our association of pink with girls stems from European legend, which professed that baby girls were born inside of pink roses.

European legend also holds that baby boys are born in blue cabbage patches—yes, the same patches that spawned the doll craze of Christmas 1983.

Why is the NBA shot clock 24 seconds?

During the 1953–1954 season, the National Basketball League was beset by difficulties. Attendance was low; many franchises were in financial trouble.

Professional basketball's problem was not a trivial one: Fans found the game boring. Hoop fans like to see plenty of shooting and scoring, but the rules did absolutely nothing to encourage teams with a lead to shoot the ball. If a team led in the late stages of the game, the custom was to have its best ball handler dribble in the backcourt, forcing opponents to foul intentionally, resulting in tedious but profitable free throws for the stalling team. There was also no incentive for teams in the lead to run cross court and set up their offense quickly, further dragging the pace of the game.

The owners knew they had a problem, but the solution was the brainchild of an unlikely savior named Danny Biasone. Biasone, a bowling alley proprietor, bought the Syracuse Nationals franchise for the princely sum of $1000. Biasone might not have had the clout within the league to compete with the Knicks or Celtics owners, but he concluded that a clock was necessary to force players to shoot at regular intervals and speed up the game.

How did Biasone arrive at 24 seconds? He figured that the average game contains about 120 shots between the two teams. Since there are 48 minutes, or 2880 seconds, in an NBA game, teams averaged exactly one shot every 24 seconds. Figuring that players would be forced to shoot before the 24 seconds expired, a shot clock would compel teams to shoot more often and, presumably, score more often.

Biasone invited club owners to watch a demonstration of how a game would be played with a clock. All could see that the shot clock would add excitement to the game, and it was instituted in regular play at the beginning of the 1954–1955 season.

The shot clock changed basketball immediately. Scoring did increase, an average of 14 points per game in one season. Most importantly, attendance rose quickly. NBA historian Charles Paikert quoted former league president Maurice Podoloff as saying that the adoption of the clock "was the most important event in the NBA and Danny Biasone is the most important man in the NBA."

Biasone's shot clock had another effect that perhaps he did

not foresee—it changed the type of player needed to build a championship team. The Minneapolis Lakers dominated the NBA before the shot clock, led by the physically bruising but slow and lumbering George Mikan. The Lakers, with the shot clock, could no longer afford to loiter downcourt while Mikan hauled down a rebound and casually jogged across the half-court line. Mikan retired the year the shot clock was instituted. He returned for the 1955–1956 season, but he averaged only 10 points versus a career average of 22 points, and he quit after half a season.

The shot clock was tailor-made for the team Red Auerbach was fashioning in Boston. In Bill Russell, the Celtics found a tall center who was also exceptionally quick and could spark a fast break offense.

Although Paikert notes that Biasone has so far been denied a place in the basketball Hall of Fame, he was justly rewarded in one respect. In the premier season of the 24-second clock, his Nationals won their first and only championship. Biasone sold the Nationals in 1963. They became the Philadelphia 76ers and went on to win many more championships.

How many more shots are taken today than in Biasone's era? In the regulation 1984–1985 season, NBA players took 168,048 shots in 943 games, an average of 178 shots per game—58 more shots per game, an almost 50 percent increase.

Dividing the number of shots per game (178) into the number of seconds per regulation game (2880), we find that a shot is taken on an average of every 16.17 seconds. Considering how many quick shots and tips are attempted on the offensive boards, which would bring down this average, it is surprising how much time most offenses take in getting off shots, and perhaps a tribute to the defensive skills in the NBA.

Why is butter, rather than margarine, served even in grungy eating establishments?

Most consumers prefer butter to margarine. But they buy margarine in grocery stores. Margarine is cheap, it has no cholesterol, and it provides a reasonable imitation of butter. One would expect to be served butter in an elegant restaurant, but why do coffee shops, cafeterias, diners, and even fast-food establishments invariably serve butter?

The answer lies in the Federal Food, Drug and Cosmetic Act, which specifically provides that oleomargarine cannot be sold in a public place unless:

> a notice that oleomargarine or margarine is served is displayed prominently and conspicuously in such place and in such manner as to render it likely to be read and understood by the ordinary individual being served in such eating place or is printed or is set forth on the menu in type or lettering not smaller than that normally used

to designate the serving of other food items. No person shall serve colored . . . margarine at a public eating place, whether or not any charge is made therefore, unless (1) each separate serving bears or is accompanied by labeling identifying it as oleomargarine or margarine, or (2) each separate serving thereof is triangular in shape.

With the exception of hospitals and kosher restaurants, few culinary institutions want to publicize the fact that they serve a perfectly wholesome substitute for butter. The restaurant industry is so self-conscious about the whole butter/margarine issue that many places serve individual pats with covers that blare B-U-T-T-E-R in capital letters.

Ironically, restaurants are under no compulsion to announce their use of margarine in cooking, where its use is much easier to conceal. And even restaurants that use butter can purchase a wide range of quality. Butter is rated on a range of 0–100, with the top butters achieving the mid-90s level. Although few restaurants use them, the cheaper, inferior grades of butter, which do not have to be labeled in any way, can have an off taste far more objectionable than any margarine.

On *Jeopardy* what is the difficulty level of the daily doubles supposed to be?

Even devout watchers of *Jeopardy* are unlikely to know the answer to this Imponderable. We watch contestants risking $300 or conservatively wagering only $40 that they can construct the right question to "answer" a daily-double answer correctly. But do they know how difficult the question is going to be?

If a daily double appears behind a $100 answer, does this

mean that the daily double will have the same difficulty as the $100 answer it replaces? Or are daily doubles more difficult? Or do they vary from answer to answer?

According to Alex Trebek, host and producer of the current incarnation of *Jeopardy*, a daily-double answer is exactly the same level of difficulty as the answer that would appear without the daily double. In fact, the staff does not even compose separate answers for daily doubles.

Although the categories under which daily doubles appear are randomly selected, faithful viewers of the show can attest to the fact that daily doubles tend to be placed in the middle range of difficulty, rarely instead of the easiest or hardest answer. In the original *Jeopardy*, contestants tended to select the easiest answers first and then move down the board neatly in ascending order of difficulty (and prize money). This worked well from the producers' point of view, since games could swing dramatically toward the end of Double Jeopardy, when more prize money was being gambled. Placing the daily double in the middle of the board helped guarantee that contestants wouldn't select them early in the game, when their appearance has a less dramatic effect on the result.

The contestants on the new *Jeopardy*, however, are more rebellious. They have taken to selecting the most difficult answers first, which makes some sense, since it assures them the opportunity to go for the largest amount of money. Usually, time elapses before all the answers can be tried—and the leftovers, from the players' standpoint, might as well be the cheap answers. Some players are being so unsymmetrical as to start in the middle of categories and work back and forward. This seems to noticeably upset Alex Trebek. He feels it is poor strategy, since contestants are thrown difficult answers before they understand the context of the categories—not all of which are totally obvious. It might also upset Trebek that the varying pattern of answer selection by contestants makes it harder to ensure that daily doubles, the wild cards of *Jeopardy*, will be selected toward the end of each round, when they will presumably help

to keep viewers pinned to their seats until the end of Final
Jeopardy.

Why doesn't a clinical thermometer register room temperature when you take it out of your medicine cabinet?

We trust thermometers. If our temperature is 98.8 degrees,
we say we have a fever. But when we take out the thermometer, the temperature reading seems to have no correlation to
reality. Why isn't the thermometer sensitive enough to know
that room temperature is much lower than 96 degrees, or whatever the lowest number on the thermometer is?

In order to understand this phenomenon, we need a crash
course in thermometer anatomy. The metal part of the thermometer that we stick into our mouths is the bulb. The rest of
the thermometer is known as the stem. The mercury flows within
a capillary, a narrow piece of glass called the mercury column.
This column is quite narrow; the mercury in the thermometer
is about the width of a human hair. At the base of the mercury
column, near the bulb (and the lowest temperature numerals),
you'll see a bump, which is called the constriction.

The constriction is the key to how a clinical thermometer
works. To create the constriction, one spot of glass is heated to
create a bump—controlled warping. The constriction works as
a physical impediment to keep mercury from going down toward
the bulb unless you shake it. If you don't shake the thermometer, the mercury will only go up, not down. The only reason
any temperature in a thermometer rises is because the mercury
in the bore of the thermometer expands. When the mercury

retracts, the constriction is large enough to stop the flow of mercury.

If you take out your household thermometer and examine its packaging, you will probably see a note that indicates that the thermometer "conforms to ASTM E667." This gibberish refers to the fact that all U.S. manufacturers of thermometers have voluntarily agreed to meet the standards of the American Society for Testing and Materials, an organization that sets standards for many products and services. ASTM is a nonprofit educational association, founded in 1898, that publishes over 7000 separate documents detailing standards in fields ranging from steel and chemicals to robotics, medical devices, and child-resistant packaging. Committees, comprised of volunteers, contribute their time to set standards, and ASTM bylaws require that a majority of committee members may *not* be comprised of producers of the item for which the standards are being set.

The ASTM specifies that clinical thermometers have constrictions, and there is no reason for the industry to want to change the technology; after all, the constriction is cheap and efficient and requires no moving or mechanical parts that could fail to keep the mercury from returning to the bulb. The ASTM standards also help explain why all thermometer scales look so much alike. All clinical thermometers are expected to have scales ranging from at least 96 to 106 degrees Fahrenheit, and graduated in 0.2-degree Fahrenheit intervals. The only long lines allowed on the temperature scale are full-degree gradations and, at the producer's option, the 98.6 degrees designation.

The ASTM also sets minimum standards for the accuracy of clinical thermometers, in degrees Fahrenheit:

Temperature Range	Maximum Error
>96.4	0.4
96.4 to 97.9	0.3
98.0 to 101.9	0.2
102.0 to 106	0.3
<106	0.4

A clinical thermometer is designed to retain the body temperature of the user until it is reset, but a thermometer will respond to hotter temperatures. Many a thermometer has been broken in the mistaken belief that it is best to rinse off the bulb by using extremely hot water.

Why do Corn Flakes and Sugar Frosted Flakes have the same number of calories per serving?

If Sugar Frosted Flakes are Corn Flakes with added sugar, how can both cereals contain 110 calories per one-ounce serving? How can Sugar Crisp, more than one-half sucrose and other sugars, contain the same 110 calories per ounce as Cheerios, which are less than 5 percent sugar?

The answer is embarrassingly simple. All cereals are composed of two kinds of carbohydrates: complex carbohydrates are found in the starchy grains used in cereals (corn, oats, wheat, etc.); simple carbohydrates are represented by the sucrose and other forms of sugar and syrup used to sweeten cereals. A presweetened cereal such as Count Chocula might have the same total carbohydrate value as the non-presweetened Corn Total, but the proportion of sugars to starches will be radically different. Kellogg's Corn Flakes and Kellogg's Sugar Frosted Flakes have the same number of calories per serving because *all carbohydrates, both simple and complex, contribute exactly four calories per gram.*

How has 110 calories become the industry standard? The average nutrition count in American dry cereals per ounce serving is:

24 grams of carbohydrate	@ 4 calories per gram =	96 calories
1 gram of fat	@ 9 calories per gram =	9 calories
1 gram of protein	@ 4 calories per gram =	4 calories
	total:	109 calories

Our average cereal, rounded off to the nearest 10 (as are all cereal labels), matches Corn Flakes' and Frosted Flakes' 110 calories.

The above chart points out another little-known piece of calorie trivia. Just as all forms of carbohydrates are four calories per gram, all proteins are four calories per gram. The protein in a greasy hamburger is no more fattening than the protein in lean halibut. Beef contains a higher proportion of fat than halibut—that is the only reason hamburger is more fattening.

There are only two ways to significantly raise or lower the calories in cereals. To add calories, add fat. General Mills's Donutz line of cereals has more than double the industry average of one gram of fat per one-ounce serving, but there is still much more fat in the whole milk that is poured into the cereal bowl.

To drop calorie counts, add fiber, which contains *no* calories. Kellogg's Most, a presweetened cereal with added fiber, is only 100 calories per ounce. Bran products and many puffed cereals with added fiber are often less than 100 calories per ounce.

The one-ounce serving referred to on the nutrition label is a measure of weight, not volume. An ounce of dry cereal often approximates one cup in volume. Because of the additional weight of the sugars, there will be less presweetened than unsweetened cereal in a one-ounce serving.

Since dietetic desserts are touted by their manufacturers as "rich" or "sweet," it is surprising that no cereal manufacturer has ever tried to market a presweetened cereal as "no higher in calories" than an austere "adult" cereal like Total or Special K.

Why do pennies and nickels have smooth edges? Why do all other U.S. coins have serrated edges?

The first generation of United States coins had smooth edges. It wasn't until the use of the steam press that it became technologically and economically feasible to create coins with reeded edges.

The serrated edges are not there for ornamentation. Back in the good old days when coins were made out of silver or gold and actually had intrinsic value, nefarious types used to pull a favorite scam. They would file or clip the edges off coins. If they were diligent in their work and had access to enough coins, they could collect the valuable silver or gold chips and then palm off the amputated coin for its face value, turning a tidy little profit.

Milled edges proved to be an excellent deterrent, safeguarding the integrity of the legal weight of the coin by making it obvious to the recipient whether or not the coin had been tampered with. If a silver dollar had a smooth edge, a banker or merchant would know that some miscreant had scraped it, and could refuse to accept it. Although many superstitious people tear corners off dollar bills, the paper itself was never purported to have intrinsic value—only the promise of the United States government to redeem it. Bearers of silver or gold coins, however, clung to the notion of today's survivalists—that even if the federal government went down the tubes, gold would still be "worth its weight in gold."

The 20-cent piece is the only United States silver or gold coin with a smooth edge. Most have reeded edges, but many early gold and silver coins have lettering (e.g., ONE HUNDRED CENTS ONE DOLLAR OR UNIT) and a few have decorative designs.

After World War II, when the supply of silver fell and its price zoomed internationally, most countries eliminated or substantially reduced the silver content in their coins and used gold only for medals and commemorative coins aimed at numismatists and investors. In 1965, the United States took this measure, and many coin collectors hoarded silver coins, figuring that *any* silver coin would soon be valuable when the country was flooded with copper-nickel coins. Their optimism was justified but premature. When speculation hit its peak, with silver selling for $50 an ounce before the Nelson Hunt fiasco, pre-1965 coins were sold like scrap metal; the price rise had rendered the small differences in value between common coins obsolete. All pre-1965 Roosevelt dimes, for example, sold for more than the most valuable circulated dime was worth before the silver craze.

Silver is now selling for slightly more than one-tenth of what it fetched less than five years ago. The silver content of pre-1965 coins still exceeds their face value, so there is no reason for anyone to deface them. But there is also no need for the reeded edges to remain on American dimes, quarters, half dollars, and "silver" dollars. The copper-nickel isn't worth clipping—would-be criminals would be better off using metal detectors at the beach. Out of custom and inertia, the reeded edges remain on American coins, a nostalgic throwback to when there was a correlation between the actual and purported value of American money.

How do they unclog mail chutes in skyscrapers?

In tall office buildings, most experienced secretaries faced with a stack of important letters will take the elevator and deposit them on the ground floor repository rather than trust them to the mail chute. Accidents will happen, and letters put in mail chutes occasionally get stuck.

When a letter gets stuck, say, on the eighth floor, it is usually visible through the glass panels located above and below the metal plate where the mail is deposited. A call to the local post office is in order. In a skyscraper-saturated city like New York, the main post office has a separate phone number just for clogged mail chutes.

Someone from the post office will arrive bearing a ring heavily laden with keys. On every floor, there is a lock on the metal plate that can be opened only by postal workers. Mail chutes have proved to be remarkably durable, so as their design has evolved over the decades (even though Cutler Mail Chute Co. has long been the largest manufacturer of mail chutes), their locks have changed and a new key is necessary. It might be necessary to try thirty or forty different keys to open up the plate.

If the blockage is visible, the postal worker simply removes the glass panel above or below the metal plate and removes the offending piece of mail. What if the blockage is between floors and thus not visible? Then the harassed businessman pictured above just might have the right idea. Chances are, the postal worker will enlist the help of the janitorial staff and use the blunt edge of a broom to nudge the blockage up or down. A broom or stick will only be used as a last resort, as any instrument can damage an envelope. About 90 percent of the time, reports Walter Koob, supervisor of collection and delivery at New York's main branch, postal workers can get the envelope out with their hands.

Koob mentioned that there seems to be an increase in mail blockages around holidays and that the biggest offenders are oversized greeting cards. If you can't easily deposit a letter in a mail chute, don't try to force it. The reason these greeting cards tend to get stuck is that lazy depositors fit them in the chute by folding them over. This is a no-no of postal etiquette. If you fold over your contemporary birthday card, you stand a decent chance of needing to buy one of those stupid "so I forgot your birthday" cards as well.

When running into the dugout from his defensive position, why is the first baseman thrown a baseball from the dugout?

Most major league baseball teams have the first baseman take custody of the ball that will be used for infield drills while their pitcher is warming up between half innings. When in the dugout while his team is at bat, the first baseman keeps the ball thrown to him in his glove.

One might expect that the catcher, the general of the infield, would be given this responsibility, but the catcher is saddled with one time-consuming fact of life alien to other infielders—in order to prepare to take the field, the catcher must don a mask, chest protector, and knee guards. The first baseman, who is the "catcher" of all the other infielders during the warm-up period (since the catcher is preoccupied with the pitcher), is thus given the not too heavy responsibility of tending to the ball and getting the infielders loosened up as soon as possible.

Why is film measured in millimeters rather than inches?

Since Kodak has always been the biggest power in film sales, it seems strange that the metric system has long ruled the film world. It turns out it wasn't always so. Kodak directly answered this imponderable:

Except for 35 millimeter film, which is really cine film, most still camera films were made in inch sizes. It is only since the Japanese became the major camera manufacturers that "two-and-a-quarter square" became 6×6 centimeters. Some of the "metric sizes" of photographic paper are only soft conversions of inch sizes. The Kodak AG catalog lists 20.3×25.4 centimeters paper—which turns out to be 8×10 inches.

R E F R E S H M N T
AL'S INFINITE CINEMAS 1·2·3·4·5·6· 12·13·1

Why have many movie theaters stopped popping their own popcorn?

The popcorn business in the United States ain't peanuts. Americans, the largest per capita consumers in the world, eat over 10 billion quarts of popcorn annually, thus generating over $1 billion for the popcorn industry.

About 70 percent of all popcorn is consumed in the home and approximately 30 percent is bought in theaters, carnivals, amusement parks, stadiums, etc. But over 75 percent of the *revenue* from sales comes from popcorn bought outside of the home. About $250 million, or around one-quarter of *all* popcorn sales, is delivered by movie theater concession sales.

To understand how crucial popcorn sales are to the movie industry, consider the economic facts of life for the movie theater exhibitor. Each owner tabulates his "nut," the total fixed costs and overhead needed to keep the theater open. In a large city, with a medium-sized house in a nice district, that nut might be about $12,000 a week. Let us assume that this theater, the Rialto, shows first-run movies and has booked the latest James

Bond thriller for the Christmas season. The owner has committed the theater to this picture many months in advance. Often, because distributors want to place their movies in houses that can run them for a long time, he might be forced to stick with an already faded movie in his theater until James Bond comes to the rescue. If *Friday the Thirteenth Part Thirteen* is grossing only $8000 a week, the owner must eat the $4000 difference between his nut and his gross.

Even James Bond does not guarantee the exhibitor endless riches, for the film distributor wants his piece of the 007 action. And it is a rather large piece. The exhibitor does not pay cash for the right to run a movie; he gives the distributor a percentage of his gross, after the nut is deducted. In the case of most first-run movies, exhibitors must pay the distributor *90 percent* of the net. If James Bond grosses $62,000 the first week, a superb showing, the exhibitor deducts the $12,000 nut from the gross (leaving $50,000), keeps a measly 10 percent, or $5000, for himself, and then sends the rest of the money to the film's distributor (usually, but not always, the company that produced the movie). By the fifth week of James Bond's run, the theater might be lucky to clear $1000 a week from the ticket receipts.

But do not cry for the theater owner. He has a secret weapon: the concession stand. Popcorn. Soft drinks. Candy. The movies may pay the bills, but the concession stands send the family to Florida in the winter.

Let's look at how concession sales affect the bottom line of the Rialto. In large cities, about 15–20 percent of all customers will stop at the concession stand (in smaller towns, even more customers eat), and the theater owner figures to gross about 75 cents for every customer who walks through the turnstile, meaning that the average purchase is over $3. The key to making money in the concession area is maintaining a high profit margin, and the items sold do a terrific job. The average profit margin on candy—77 percent; on popcorn—86 percent; on soft drinks—a whopping 90 percent. For every dollar spent at the concession counter, the theater operator nets over 85 cents.

This is the theater's average cost for a large bucket of "buttered" popcorn that might retail for two to three dollars:

Popcorn— 5 cents
Butter Substitute— 2 cents
Bucket—25 cents

Yes, the bucket itself is the most expensive component of your popcorn purchase. Even if the Rialto were to use "real" butter, which most consumers can't distinguish from imitation, it would only add three more cents to the cost.

Remember that the Rialto has netted $5000 from the admissions to the first week of the James Bond movie. But it will gross almost $10,000 and will net over $8000 from the concession stand. And on the fifth week, when the Rialto nets only about $1000 from admissions, it will earn almost $3000 extra from food and drink sales.

Considering the importance of popcorn, the largest grossing concession item in profits, why would exhibitors deny the tradition of popping their own corn? Even in the 1980s, a good majority of theaters still pop their own. Many exhibitors believe that popping their own corn adds luster to what is an impulse item. The sound of the popping and the aroma of fresh corn and (usually) fresh oil is tantalizing to the vulnerable. And it is slightly cheaper for theaters to buy kernels rather than purchase already popped corn from a food distributor.

But the crucial question remains: Does on-site popping increase sales? A growing number of concession experts at the big movie chains believe that there is no evidence that on-site popping affects purchases one way or the other. Most of the big chains do not have a strict policy at all on the question. While one theater chain, Walter Reade, told *Imponderables* that its sales are higher in sites with on-premise popping, a representative from Loew's disagreed strongly, arguing that none of the research and none of Loew's internal experiments support the contention that consumers are driven into even a frenzy-ette by their proximity to exploding kernels.

There are plenty of reasons why managers dislike on-premise popping. Equipment can get messy and smelly, offending both workers and potential customers. Poppers can also break down, and as simple as it may sound, managers must constantly train high-turnover employees how not to wreck the equipment. Commercial prepopped corn is uniform in size and taste, whereas homemade popcorn is subject to the vagaries of oil temperature and stubborn kernels refusing to pop. Most important, theaters never run out of prepopped corn. No manager wants to see his sales force frantically loading the popper while customers wait impatiently in line, contemplating bolting for the theater.

If there were much consumer resistance to prepopped corn, you would see machines in every theater lobby in America. But the quality of packaged corn can be as good as fresh-popped. The crucial element in consumer acceptance of popcorn is its moisture content. Moisture is the enemy of popcorn, and "old" popcorn can be restored by being placed in the heating chambers that virtually every concession stand possesses. Left at room temperature, popcorn reabsorbs moisture from the atmosphere. In the warmer, at an ideal 135–155 degrees, the moisture is driven out. The lesser moisture in theater popcorn is what makes it taste better than its packaged, unheated counterpart found at supermarkets or ball games.

If concessions are the crucial moneymaker for theaters, why aren't stands more adventurous in their offerings, and why don't they offer more choices?

The key to this answer is a favorite word of all food purveyors—*turnover*. The theater owner wants to be able to process as many customers as possible in a short period of time. Now that double features have become a thing of the past for most theaters, concession stands must brace themselves for an onslaught of customers arriving at approximately the same time. More than 80 percent of all concession sales are completed immediately after the ticket purchase, before the customer has taken a seat in the theater. Nothing will turn off a potential customer more than long lines at the food stand. The fewer choices a customer has to make, the less anxiety the customer feels and, most important, the *faster* the customer is likely to decide what to buy. Most concessionaires have found that when they introduce new products, such as chocolate chip cookies or frozen yogurt, it eats into the share of money that their old products would have garnered but does not generate additional revenue or attract patrons who didn't previously buy food at the theater.

Loew's has found that by decreasing its number of food and drink options, it can generate faster turnover without causing any consumer resistance. It purposely does not emphasize its candy display and provides only 12 options, since it has found that any more choices only tend to befuddle the customer and slow down his or her decision-making process. Even so, candy provides about 20 percent of Loew's concession sales.

Hot dogs provide about the same profit margin as popcorn, but their gross sales are minuscule in comparison. Hot dogs are provided partly as a meal substitute for those taking in a movie at the lunch or dinner hour. Hot dogs are problematic because, unlike popcorn, they can't be resold the next day. Concession stands can sell only a limited number of hot dogs efficiently. With the rotogriller, the horizontal contraption with rotating silver tubes, hot dogs cook quickly but can shrink during movies considerably shorter than *Lawrence of Arabia*. The pinwheel cooker, the ferris wheel arrangement where spiked hot dogs rotate over a heating element on the bottom, cooks fewer dogs more slowly but at least doesn't turn quarter-pounders into cocktail franks.

Frozen desserts are a particular bane to concessionaires. They represent about only 1 percent of sales. Bon Bons are the biggest frozen item only because the theaters can sell them with a hefty profit margin. If theater owners tried to sell gourmet ice cream cones, they would have to charge several dollars a scoop to maintain their profit margin, and then pay for it in messier theater floors. Freezer cases are particularly vulnerable to employee ineptitude. If a worker turns off the freezer switch by mistake, profits melt along with the ice cream.

If brussels sprouts would sell in theaters, concessionaires would find a way to cook and sell them. American audiences simply reject the attempt to foist any products other than the big three (popcorn, soft drinks, candy) on them. The concessionaire merely responds to what the consumer wants. If we really cared that popcorn be popped at the theater itself, it would be.

How do the networks sell advertising time when live programs run longer than scheduled?

Who knows how long the Academy Awards will last? Or the Super Bowl? Certainly, the networks don't. With thirty-second commercial spots fetching hundreds of thousands of dollars, you can be assured that big money is at stake. Obviously, networks would like to sell commercials during overruns, but how can they sell time when they don't know if they are going to have it? And what about the local affiliate, which usually airs its own ads at 11:00 P.M., when the Academy Awards is just getting to the important nominations?

When they air an event that they know has the potential to run past its allotted time, the networks try to sell advertising spots on a contingency basis. ABC might approach Kraft and say: "Do you want to buy a spot on the Oscars after the third hour?" Kraft would argue that ABC can't guarantee placement of the ad (after all, in 1985, the Oscar broadcast almost came in on time). ABC would counter with a reduced price—something on the order of a 30 percent discount—to compensate Kraft for the possibility that the commercial will not air. ABC is happy that it has eked out some gravy for commercial time it would have otherwise not sold. Kraft is happy because it gets a bargain rate and reaches an audience likely to hang in to find out who won for best picture. Likewise, overrun time on sports programming is likely to be a bargain: Viewers will stay tuned to see who wins the contest, and afternoon events that run long tend to bleed into prime time (in some parts of the country, at least), when the number of sets in use is higher.

Why don't sponsors jockey to buy overrun time? For the most part, commercials are bought by advertising agencies representing sponsors. Commercials are usually designed to influence specific demographic groups, and advertising time is bought in order to reach a designated number of that group within a certain amount of time. Sponsors tend not to be as concerned about "bargains" (they know approximately how much it will cost them to reach each thousand of their targeted audience) as they are about reaching that audience efficiently (they don't want to sell life insurance on *Falcon Crest,* whose audience is predominantly female when most of their customers are male) and quickly.

Many companies use live programs (sports, awards shows) that might overrun to introduce new products, announce improvements and changes in image of products, since specials and sports are exciting and glamorous environments in which to showcase their "exciting news." When a company is making such an important announcement, it is imperative that commercials run as scheduled, to coincide with its products' hitting the stores.

Networks aren't always successful at selling overrun time, however. If not, their best strategy is to use the vacant advertising time to promote their own shows. Ever since ABC used the 1976 Olympics to successfully hype its prime-time line-up for the fall, networks have become acutely aware of the power of promotion within important television events to increase the initial tune-in of regular series. ABC used the same tactic with the 1984 Olympics to promote *Call to Glory.* The operation was successful (the pilot received a huge rating), but the patient (and show) died.

The last option of the network, and by far the least desirable, is to use up the extra commercial time by running free ads. When networks haven't sold time and haven't planned extra promo time, they will often run ads at no cost to the sponsor rather than run public-service spots. Public-service spots denote to the viewer that no commercial time *could* be sold, a failing the networks do not want conveyed, even subliminally, to the viewer.

When network overruns impinge on their affiliates' time (11:00 P.M. E.S.T., 10:00 C.S.T.), the local station usually loses the revenue from commercials already sold for that period. In most cases, local stations sell time in "strips," meaning that sponsors buy, say, five 30-second spots during the 11:00–11:30 P.M. period, Monday through Friday. The station may place the sponsor's five spots on whatever day or days it wishes to. If the network preempts its time, the local station will simply place the ad on another day. If the station were totally sold out of commercial time for the quarter and the network preempted it, the station may have to refund the sponsor's money unless some kind of trade of time slots can be negotiated. It thus isn't hard to understand why local affiliates don't appreciate even planned overruns, such as theatrical movies that are longer than two hours. Although local stations profit from the limited number of commercials they can sell per hour during the network line-up, they can make more during local programming, when the network doesn't have its finger in their pie.

Why are U.S. elections held on Tuesday?

Reformers are calling for weekend elections in order to increase voter turnout. Before we take precipitous action, perhaps it would be best to study why our founding fathers, with due deliberation, chose this day for citizens to exercise their most precious duty.

Nah. History won't help us a bit, except to surmise that

the selection of Tuesday as Election Day was just a cosmic accident. Professor Robert J. Dinkin's studies on voting in provincial and revolutionary America reflect the haphazard beginnings of elections in the fledgling democracy.

In provincial America, there was no single standard date for balloting in the thirteen colonies. Voters had to travel to their county seat to mark a ballot; smaller precincts did not yet exist. Most elections in the northern colonies were conducted in mid-spring or early fall so that snow wouldn't prevent far-flung voters from arriving in time, since a trek of twenty-five miles or so by horseback was often necessary to reach polling places. The larger colonies would often allow voters more than one day to vote. Dinkin observes in *Voting in Provincial America* that "Maryland allowed the polls to stay open as long as the candidates could show cause that additional voters were en route, even if it took a number of days. Heated contests in Baltimore, Dorchester and Frederick counties thus frequently lasted three or four days."

Lest we imagine that our forefathers were much more pious than we, willing to embark upon long voyages in order to do their part for the common good, it must be added that voting was often more of a social occasion than a civic one. Election Day was a big to-do, with drinking and carousing the order. Crowd control was a major problem, since provincial capitals and county seats did not have the facilities to handle the influx of visitors.

After the Revolutionary War, election dates within each state became more standardized, but there was little uniformity among different states. Monday and Tuesday seemed to be the most popular days, but elections continued to be held mostly in spring and autumn. After 1776, most states instituted more polling places so that distant voters need not travel to county seats. Localities still had the right to leave polls open for more than one day and to set their own hours for the polls to remain open.

The first Tuesday of November was established as the date for presidential elections prior to the election of 1848. Many states still conducted their elections on Monday. Dinkin mentions that the phrase "As Maine goes, so goes the nation" stems from the fact that Maine's elections were held on the second Monday of September; along with Indiana, Ohio, and Pennsylvania's October elections, Maine's voters were thought to provide a barometer of public opinion that might foretell trends relevant to the presidential election in November.

The importance of these October elections was so manifest that these states were infested by national party leaders who used greenbacks and arm-twisting tactics to cajole midwestern voters into seeing things their way. By the mid-1880s, the political pressure reached such a pitch that the "October states" gave up their early contests. Although Maine held on to its Monday-in-September elections until 1949, most states saw the wisdom in uniting state and national elections, and despite the few diehards, states chose to conform to the federal election choice of the second Tuesday in November.

Why do people look up when thinking?

Medical doctors have a nasty habit. You pose them a particularly tough Imponderable and they answer, "I don't know." Most medical and scientific research is done on topics that seem likely to yield results that can actually help clinicians with everyday problems. Determining why people look up when thinking doesn't seem to be a matter of earth-shattering priority.

Ironically, some serious psychologists *have* decided that this question is important, have found what they think is a solution to the Imponderable and, most amazingly, found a very practical application for this information. These psychologists are known as neurolinguists.

Neurolinguists believe that many of our problems in human interaction stem from listeners not understanding the frame of reference of the people speaking to them. Neurolinguists have found that most people tend to view life largely through one dominant sense—usually sight, hearing, or touching. There are many clues to the sensory orientation of a person, the most obvious being his or her choice of words in explaining thoughts and feelings. Two people with varying sensory orientations might use totally different verbs, adjectives, and adverbs to describe exactly the same meaning. For example, a hearing-oriented person might say, "I hear what you're saying, but I don't like the sound of your voice." The visually oriented person might say, "I see what you mean, but I think your real attitude is crystal clear." The touch-dominant person (neurolinguists call them kinesthetics) would be more likely to say, "I feel good about what you are saying, but your words seem out of touch with your real attitude."

Neurolinguistically trained psychologists have found that they can better understand and assist clients once they have determined the client's dominant sense (what they call the client's representational system). All three of the above quotes meant the same thing: "I understand you, but your words belie your true emotions." Neurolinguists adapt their choice of words to the representational system of the client, and they have found that it has been a boon to establishing client trust and to creating a verbal shorthand between psychologist and patient. Any feeling that can be expressed visually can be expressed kinesthetically or auditorily as well, so the psychologist merely comes to the patient rather than having the patient come to the psychologist—it helps eliminate language itself as a barrier to communication.

When grappling with finding the answer to a question, most people use one of the three dominant senses to seek the solution. If you ask people what their home phone number was when they were twelve years old, three different people might use the three different dominant senses of vision, hearing, and feeling. One might try to picture an image of the phone dial; one might try to remember the sound of the seven digits, as learned by rote as a small child; and the last may try to recall the feeling of dialing that phone number. Notice that all three people were trying to remember an image, sound, or feeling from the past. But some thoughts involve creating new images, sounds, or feelings. Neurolinguists found they could determine both the operative representational system of their clients and whether they were constructing new images or remembering old ones before the clients even opened their mouths—by observing their eye movements.

These eye movements have now been codified. There are seven basic types of eye movements, each of which corresponds to the use of a particular sensory apparatus. Please note that these "visual accessing cues" are for the average right-handed person; left-handers' eyes ordinarily move to the opposite side. Also, "left-right" designations indicate the direction from the point of view of the observer.

Direction	*Thought Process*
up-right	visually remembered images
up-left	visually constructing [new] images
straight-right	auditory remembered sounds or words
straight-left	auditory constructed [new] sounds or words
down-right	auditory sounds or words (often what is called an "inner dialog")
down-left	kinesthetic feelings (which can include smell or taste)

There is one more type of movement, or better, nonmovement. You may ask someone a question and he will look straight ahead with no movement and with eyes glazed and defocused. This means that he is visually accessing information.

Try this on your friends. It works. There *are* more exceptions and complications, and this is an admittedly simplistic summary of the neurolinguists' methodology. For example, if you ask someone to describe his first bicycle, you would expect an upward-right movement as the person tries to remember how the bike looked. If, however, the person imagines the bike as sitting in the bowling alley where you are now sitting, the eyes might move up-left, as your friend is constructing a new image with an old object. The best way to find out is to ask your friend how he tried to conjure up the answer.

Neurolinguistics is still a new and largely untested field, but it is fascinating. Most of the information in this chapter was borrowed from the work of Richard Bandler and John Grinder. If you'd like to learn more about the subject, we'd recommend their book *frogs into Princes* (sic).

To get back to the original Imponderable—why do people tend to look up when thinking? The answer seems to be, and it is confirmed by our experiments with friends, that most of us, a good part of the time, try to answer questions by visualizing the answers.

Why do records spin at 33⅓, 45, and 78 R.P.M.?

It sounds like some kind of Polish joke, but until Emile Berliner developed the gramophone and disc recording at the end of the nineteenth century, artists had to assemble and re-create their performance every time they wanted to issue a record.

Many professional musicians weren't willing to make records before the gramophone—they couldn't afford the time or money wasted immortalizing performances that they could be rewarded handsomely for in concert. Most of the earliest records were experiments performed by amateurs—the aural equivalent of home movies. Consumers bought records at first not to hear their favorite performers, but to listen to the novelty of the human voice or musical instruments come through a box in their living room.

Berliner's gramophone proved vastly superior to its predecessors, and not only because the gramophone produced more faithful fidelity. More importantly, gramophone records were dubbed from a master disc. With the master, any number of copies of one performance could be replicated, the performers were free to spend most of their working time in front of an audience, and assembly line-like efficiency was assured in the recording industry.

The early Berliner gramophone came in three models, the cheapest of which, the Seven-Inch Hand Gramophone, sold for a little over ten dollars, much less than the earlier cylinder phonographs. Berliner wanted to have records spin at one hundred revolutions per minute, but found he couldn't get enough music on each disc. He experimented with 40 R.P.M., but the sound was awful, so he compromised at 70 R.P.M., which quickly became the industry standard and stayed so until the 1920s.

Of course, these early gramophones were hand-cranked, so 70 R.P.M. was a suggested speed, a target. The deftness and durability of the owner's wrist determined whether that speed was attained and sustained.

In 1925, a synchronous motor was attached to turntables, allowing for uniform speed and rendering the hand crank obsolete. These motors turned at 3600 revolutions per minute. With the 46:1 reduction gear standard on gramophones, the resulting speed was a little more than 78.26 R.P.M.s. The 78 was born. *These gear ratios, not any grand design or theoretical benefit, were responsible for the numbers 33 1/3, 45, and 78.*

Long-play records were invented in the 1920s by Bell Laboratories but weren't aimed at the consumer market. LPs were used as sound for motion pictures, and later by Muzak and its competitors for background music. Long-play records were a boon for "live" radio programs, enabling them to be heard at the same time in different time zones (actually, the masters were replayed separately for each time zone).

But early LPs played more music than 78's not only because they played at a lesser speed, but because they were bigger in size. Columbia Records engineer Dr. Peter Goldmark created the breakthrough necessary to propel the LP into prominence—the microgroove, which allowed a much thinner stylus to track more music on less disc. One LP (rather than five 78's) could capture a whole symphony with superior fidelity. Goldmark's streamlined phonograph, with its magnetic pickup, eliminated the need to gather energy from the vibrating needle undulating on the (now obsolete) larger grooves. The vibrating needle, a necessity on the original Gramophone, was a major reason why records spun at 70 and 78 R.P.M. in their infancy. With microgrooves and magnetic pickup, it was no longer true that (as is still true with audio tape) the faster the speed, the better the fidelity.

Although Columbia offered its new system to RCA, its major competitor (at a price, of course), General Sarnoff countered with his own system, the 45 R.P.M. record and phonograph, also employing microgrooves. But since the 45 was seven inches in diameter, rather than the LP's twelve inches, the 45 didn't offer more than four or five minutes of music.

Another Imponderable about the record industry—why do 45 R.P.M.s have a big hole, necessitating different equipment than the LP?—is quickly answered by this war between the RCA and Columbia systems. The big hole and the different speed were introduced to force consumers to buy RCA hardware. There was and is nothing mystically perfect about the 45 R.P.M. speed, nothing advantageous about the big hole save its greater convenience for automatic record changers. What *was* perfect about the 45 was that no other record player could reproduce the work of RCA's formidable roster of artists. RCA was gambling that consumers would buy new, incompatible hardware in order to listen to the right software.

The original RCA 45 R.P.M. record player couldn't play either 33 or 78 R.P.M. records, and at first, no other machines could play RCA's 45's. Consumers responded to the 45/LP battle by not buying many of the new microgroove records or their players. Similar consumer bewilderment about incompatible hardware stifled the early growth of videotapes and video recorders.

In 1950, RCA capitulated and produced its own microgroove LPs; Columbia soon issued 45's. Both formats prospered because the two discs weren't direct competitors. The 45's were perfect for popular songs, their price was right, and their size suited the jukebox industry (the only reason 45's still retain the big hole in the 1980s is so that they won't render current jukeboxes obsolete). LPs were the perfect vehicle for long symphonic works and operas, and their length led to popular works created specifically for the form (e.g., rock concept albums such as *Sgt. Pepper's Lonely Hearts Club Band* and greatest-hits collections).

All three of the dominant playing speeds for records were not predesigned—they were the byproducts of then current technology and marketing necessities. If inventors like Emile Berliner and Dr. Goldmark ran the music industry today, all records would probably play at 33⅓ R.P.M. There is no logical reason why we need more than one speed today.

A sharp-eyed reader, James D. Gibbs, a physics teacher and hi-fi and stereo enthusiast, wrote us to argue that the selection of 33 R.P.M., unlike the other speeds, wasn't totally arbitrary. He's right.

The original patent for the 33 R.P.M. noted that a minimum satisfactory linear velocity of the record groove as it moves past the stylus is necessary to produce good sound. Obviously, though, the slower the velocity, the longer the recording time. The combination of the microgroove and the slow R.P.M. enabled the LP to play for approximately 11 minutes, to correspond to the running time of a 1000-foot roll of film. When LPs hit the consumer market, their greater capacity without sacrifice of good fidelity proved to be irresistible.

Why do women wear such uncomfortable shoes?

One of the world's great Imponderables is why women are willing to subject their feet to the torture of pointy-toed high-heeled shoes. As with most great Imponderables, this is a mystery precisely because the answer is difficult to nail down. We tried our best. We contacted shoe manufacturers and designers, fashion consultants and academics. Almost as one, they had a one-word answer to the question—*vanity*.

Imponderables finds this answer a tad simplistic and condescending. Many women will go to great pains to hide the length and girth of their feet in too-small shoes, but surely there must be a better explanation for a phenomenon that has existed in Western culture for over a thousand years. After all, other uncomfortable fashion imperatives, like bustles and girdles, which became popular on waves of vanity, have bitten the dust while excruciating shoes have stayed the course. What gives?

Whenever *Imponderables* wants to get answers of considerably more than one word, we turn to psychology, which never fails to supply useful quotes on any conceivable subject. And a very eminent psychologist, Lawrence Langner, ruminated about this very Imponderable in his book *The Importance of Wearing Clothes*. Langner's thesis, to put it simply, is that primitive culture was obsessed with the actual exhibition of genitals. Clothes became one of the main ways in which a more "civilized" culture could sublimate this primitive drive. Garments became, for the wearer, a symbolization of the genitals. The farther away the clothing was from the actual genitals, the more unconscious the symbolization became. For example, cod pieces, while covering up male genitalia, obviously drew attention to the very organs they were designed to conceal. Langner argues that the feet, because they are so far from the genitals, are the perfect part of the body for genital displacement.

Langner cites several examples to show how consistently irrational shoe design has been. In the Middle Ages, a long-toed shoe or boot called the pontaine was all the rage. It was the shape of a phallus, with a snouted toe, and sometimes extended twelve full inches beyond the foot. Although it caused an outrage from pious types, it was the height of fashion. In the fourteenth century, pontaines were made even more elaborate, elongated into animal shapes, such as a bird's claw or an eagle's beak. Some were even in the shape of a phallus. Pope Urban V and Charles V of France issued proclamations decrying the shoes but were incapable of stopping the fad.

To this day, despite some women's insecurities about having big feet, sleek, long, pointed high-heeled shoes are considered sexy, immodest, aggressive, and provocative. Langner believed that a high-heeled shoe ". . . enables a woman to increase height and make her feel more attractive and sexually more interesting." In pornographic movies, it is not uncommon for female characters to peel off all of their clothes at will—but you couldn't pry away their spiked high heels with a crowbar. There may be a small audience of foot fetishists, but these shoes remain on the porn queens primarily because they have become a symbol of lasciviousness.

Chinese women have traditionally worn sandals wide at the

toes (after all, the girth of the toe area is one of the widest of any part of the foot), but Western women seem to resist common sense. As J. C. Flugel says in his book *The Psychology of Clothes*, ". . . there is still manifest a desire for a greater pointedness than is warranted by the shape of the foot, and in which objections to the unnatural shape have become almost entirely rationalized as motives of hygiene."

So maybe Langner's thesis makes some sense. It is not clear *why* a woman would want to walk around with a phallic symbol on her foot. But then it isn't clear why a woman would want to walk around with those excruciating shoes on her feet either.

Is there any difference between toffee and caramels?

Not much. They share the same ingredients: sugars (usually including dextrose and corn syrup); milk; butter and other vegetable fats. Toffee has a higher butter and cream content than caramels.

The major difference between the two candies is that the toffee is processed to a higher temperature than caramels. Not only does this higher temperature make toffee harder than caramel, it helps create the distinctive toffee taste.

What's the difference in length between a size-6 shoe and a size-7 shoe?
What's the difference in width between a size-A and a size-B shoe?

Shoe sizes have had a less than orderly history. A size 8 today wasn't a size 8 a hundred years ago. A size 8 in France today will fit an infant in the United States. Actually, sophisticated sizing systems have been in existence for only a little over 100 years, but their antecedents have existed for centuries.

We have always tended to base our measurements on parts of the human body. The inch was supposed to approximate the width of a thumb; the foot was the length of a man's foot; a yard was the length from the shoulder to the fingertips of an arm fully extended. Although many ancient cultures developed measurements for long distances, exact measurement of smaller units was a problem. It was easy, for example, to say that a foot should be the length of a man's foot, but which man's foot? Supposedly, Charlemagne's foot was the model for our foot, but other cultures use different feet to reflect their smaller stature.

Imagine the problems determining lengths smaller than one foot without the aid of precise measuring instruments. Many cultures used seeds of cultivated grains to measure short distances, but most grains are no more uniform in length than men's feet. The barleycorn, however, was one exception, with its seeds of surprisingly consistent length. The Roman system of counting was based on 12, rather than 10, so it was natural for them to seek some subdivision of a foot divisible by 12. The inch became that smaller unit, and its popularity was ensured when it was found that three barleycorns placed end to end equaled one inch. There is controversy about who first decreed that a foot should equal 36 barleycorns, but whoever did (King Edward II of England issued such an order in 1324—the question is whether someone else did earlier) forever influenced shoe sizing.

Even before the Middle Ages, some English shoemakers started to use the barleycorn standard to measure the foot, and there was a consensus that one barleycorn, or ⅓ inch, would be the logical increment separating shoe lengths; but for the most part, chaos reigned in sizing. Most people made their own crude shoes, and until the nineteenth century, those affluent enough to buy shoes had them custom-made. Early shoemakers actually had a vested interest in not making shoe sizes uniform. Once a shoemaker made a last to fit the measurement of his customer, he kept it in his possession in order to ensure that the patron came back. Since the customer absorbed the price of constructing a new last, the shoemaker usually snared repeat customers.

There were a few false starts in developing a sophisticated sizing system for shoes, but in 1880, an American, Edwin B. Simpson, introduced the system we use today. Although it included the English increment of ⅓ inch per each full size, Simpson's system was a breakthrough in many ways:

- It introduced width sizes. (Width sizes, then and now, measure the girth of the balls of the feet.)

- It introduced proportional measurement. Up until the Simpson system, no accommodation for other measurements was made when the length of a shoe was increased. Simpson increased the size of the ball, waist, instep, and heel as overall last length increased. This adjustment made mass production of shoes possible.

- It introduced half sizes (which are ⅙ of an inch).

- It established separate sizing systems for infants', children's, men's, and women's shoes.

Simpson's system did not meet with immediate acceptance, however, for its half sizes and width differentiations forced manufacturers to make many more models of each type of shoe and forced retailers to stock a much greater inventory. But the superiority of the system, particularly the better fit it provided for customers, forced the United States industry to adopt it as its standard.

How did they figure out how long a size 1 would be (a size 1 is clearly longer than ⅓ of an inch)? The genesis of children's and adult sizes is murky, but the best explanation we've seen is found in a manual called *Professional Shoe Fitting* (published by the National Shoe Retailers Association and written by William A. Rossi and Ross Tennant), which also supplied us with most of the technical information for this Imponderable. According to Rossi and Tennant, it is a probably a fluke that children's sizes top at 13½ and then start again with an adult 1. The earliest English shoemakers used four inches as their smallest shoe size because four inches was the approximate length of the infant's foot when it was first ready for shoes, and shoemakers in pre-ruler days were used to measuring four inches by noting the span of the knuckles across the hand (try it, it still works). Then, the theory goes, they decided to end the children's sizes, somewhat arbitrarily, by adding the approximate span of a hand (nine inches) to the original four inches, and arrived at the number thirteen. In this case, of course, the added nine "inches" were actually nine one-thirds of an inch—since few adults, let alone children, have feet thirteen inches long.

Another theory postulates that since the English adopted the Roman counting system, based on 12, rather than the metric system, shoemakers used measuring instruments with twelve markers (a 12-inch ruler actually has thirteen inch markers), shoemakers might have decided to go with 13 instead of the even 12 (remember that the English did not use half sizes at this time, and hence did not need utensils to measure half sizes). The only problem with this theory is that shoemakers weren't using inch rulers to size shoes, but rather worked in one-third or one-half inch increments.

Until Simpson's system, shoes, if they came in widths at all, were offered in only two sizes—fat and slim. Simpson's system for width was just as precise as his length designations, and they are still used today. *For every additional width size, there is a one-quarter inch in circumference or girth added around the ball on the last.* But the particular genius of Simpson's system was to automatically increase girth measurement as length size increased. For each full-length size, an additional one-quarter inch of circumference is added to the width (e.g., a man's 8D size will be one-half inch wider than a 7C; an 8B will be the same width as a 9A). Simpson's width system forced retailers to literally quadruple their inventory, so it wasn't until well into the twentieth century that most stores stocked shoes in all of these sizes. Even today, particularly with leisure and athletic shoes, manufacturers and retailers are reluctant to offer a full range of width sizes. Some shoe manufacturers issue widths in only three sizes, narrow, medium, and wide. There are no official boundaries for these sizes, but Rossi and Tennant indicate that in men's sizes, a narrow is between a B and a C, a medium between a C and a D, and a wide somewhere between a D and an E.

Any shoe retailer who wants to stock all of the "normal" sizes would be obliged to carry nearly three hundred different combinations. This proliferation of sizes places the retailer in a classical marketing dilemma. The shoe seller can choose to stock fewer styles with the ability to fit more customers or to increase the range of choices but risk not being able to fit patrons with particularly long, short, fat, or thin feet. Still, by carrying only the fifteen most popular size combinations of shoes for both men and women, a salesperson can fit shoes for about two-thirds of the American public.

The most popular sizes for women are 6½AA–9AA and 5B–9B. The single most popular woman's shoe is the 7½B, which represents almost 7 percent of all shoe sales. Almost one-half of all women take a B width, and less than 1 percent wear the extreme widths of AAAAA, E or EE. Less than 1 percent of all women wear sizes less than 4 or more than 11; 7½ narrowly beats out 7 for favorite length; over three-quarters of all women wear sizes between 6 and 9.

The 8½D and 9D are tied for the most popular male size, both with an 8 percent share of the market. The fifteen most popular sizes for men's shoes are 8C–10C and 6½D–11D. Over 60 percent of all men wear D widths, but the distribution of other widths is more spread out than women's, with B, C, and E all over 7 percent. Size 9 barely beats out 8½ as the most popular men's length, as any bowling alley proprietor could tell you.

Feet have gotten progressively larger over the last few centuries, and at an accelerated rate in this century. According to Rossi and Tennant, the average shoe size of soldiers in the Revolutionary War was 6C; in World War II, 8D; and in 1984, 9½D. The adolescent boy or girl of today is likely to wear shoes as large as those of the parent of the same sex and much larger than those of the grandparent.

You will sometimes find shoes without the usual one- or two-digit size listed on the inner lining. These shoes probably utilize the "standard system," which is nothing but a code for retailers to let them know what the exact size of the shoe is without letting the customer know. In the standard system three-digit code, the first number indicates width (1 = A, 3 = C), the second number reveals the usual length size, and the third number confirms whether the length is a full size or half size (0 = full size, 5 = half-size). A 375 shoe would be a 7½C. Although use of the standard system has lessened, particularly with the advent of self-service shoe stores (where customers must be able to identify what size the shoes are), it still remains—a remnant of the time when shoe stores lied about what sizes they carried when they couldn't stock sufficient inventory.

European shoes not made expressly for the American market generally use French sizing, which is based on the metric system. Each increment of length is two-thirds of a centimeter, approximately one-quarter of an inch; their smaller increment eliminates the need for half sizes. The French do not discriminate between children's and adults' sizes, however, so an American men's 9 would be a 43 in the French scale.

Why don't footwear makers simply collaborate and standardize shoe sizes, so that a 9E in a Bally loafer will fit the same foot as a Nike training shoe? It's impossible. The style of the shoe makes a tremendous difference. Is it a high heel or a flat? Is it an oxford or a pump? Does the material of the shoe stretch? What shape is the last? All of these factors profoundly influence whether or not a shoe fits well.

The notion that we can fit shoes merely on the basis of length and the width of the ball is clearly silly in the first place. Someone with an extremely high arch, for example, is going to have trouble with a fit in any ready-made shoe, unless the style of the shoe happens to compensate for the idiosyncrasy. Anytime the contours of a foot do not happen to correspond to the automatic width adjustments made for the average foot, a length and width size will only begin to indicate whether the shoe will be comfortable. A jeans manufacturer and a bra-maker have both based their national advertising campaigns on the notion that folks with the same size cannot necessarily wear the same garment: The principle with shoes is the same. The volume necessary to hold a fleshy foot is simply greater than that for a shoe designed for someone with bony feet.

It's a losing fight. You might as well be resigned to a lifetime of trying on shoes. But buck up. You are now an educated consumer.

Which fruits are in Juicy Fruit chewing gum?

Imponderables paraded our wiles at the folks at the William Wrigley Jr. Company, but we weren't able to pry away from them the secret formula to Juicy Fruit gum. There are artificial as well as natural flavorings in Juicy Fruit, and Wrigley is understandably not enthusiastic about revealing their proportions, since no other manufacturer has hit on a marketable knock-off.

Still, a representative from Wrigley's was kind enough to list the predominant fruit flavorings in Juicy Fruit. They are: lemon, orange, pineapple, and banana. Perhaps the banana is responsible for Juicy Fruit's inimitable richness.

What is the difference between an Introduction, a Foreword, and a Preface of a book?

These three terms have become virtually interchangeable. One can encounter all or none of these three features in any given book, and all or none of them might be written by the author.

Traditionally, however, there has been a distinction between the introduction and the other two elements. While a preface or foreword usually tells the reader what to expect, the introduction typically starts the process of orienting the reader to the subject matter itself.

In a preface or foreword, an author might explain what burst of inspiration ignited the masterpiece you are reading. He might talk about how this book should totally change your life as you know it and about how, although his book will make you a perfect person, he is not legally or morally responsible for that transformation. He will also wittily acknowledge all of the little people whom he trampled upon in order to purvey his deathless prose. In the introduction, the author dips into the actual subject matter, supplementing what is in the book and ensuring that the reader adopts the properly respectful attitude toward his material.

Although most publishers observe the above distinction, they have varying policies about just how interchangeable the foreword and preface are. Some publishers arbitrarily title remarks by the author as the foreword and those by editors or outside endorsers as the preface. The esteemed publisher of this book labels the first prefatory passage, whether by the author or an outside source, as the foreword. If there is a second preliminary passage, it is deemed the preface. Thus, at William Morrow, you will never find a book with a preface that doesn't also contain a foreword. Nowhere has *Imponderables* found any legitimate distinction between the contents of a foreword and a preface.

Publishers do concur on the order in which these three elements should be placed in a book. The copyeditor's bible, *Words into Type*, recommends that the preface be placed after the table of contents (and after the list of illustrations, if there is one). If there are two prefaces, the editor's preface is placed before the author's. The foreword comes next. The introduction can be part of the text; if not, it comes after the foreword.

How can amputees feel sensations in limbs that have been severed?

Most amputees experience "phantom limb" sensations. Many patients report feelings as vivid and sensitive in the severed limb as the real counterpart. A patient, for example, with an arm amputated at the elbow might feel she could wave her hand, make a fist, or raise a finger. The phantom limb may feel hot, cold, wet, itchy, or painful. The most common report is that the amputee feels a mild tingling sensation or tightness in the phantom.

In most cases, phantoms start when the patient regains consciousness after surgery, but the duration of phantom limbs varies dramatically. While some patients lose feeling in their phantom limbs in a matter of months, others never lose theirs, although in most cases phantom limb sensations become progressively less distinct over time.

Sensations in the proximal parts (e.g., upper arm, thigh) tend to disappear first, with the extremities (fingers and toes) tending to linger. The amputee often perceives the (phantom) extremities moving closer to his or her stump.

Phantom limbs can occur with other forms of amputation. Some women experience phantoms after mastectomy; plastic surgeons report an occasional phantom after removal of fat or even after "nose jobs."

There are both physical and psychological explanations for the phantom limb phenomenon, with very few practitioners doubting the importance of either factor. The best argument for the organic etiology of the phantom limb is that its existence is almost universal among amputees and that there is no evidence that sufferers of phantom limb have any different psychological profile than those who don't experience it.

Most theories attribute phantoms to the sensory cerebral cortex. There is ample evidence for this supposition. The parts of the phantom most vividly felt by amputees are the digits of the hands and feet (particularly the big toe and thumb), the areas with the most representation in the cerebral cortex. The proximal areas of phantom limbs, such as thighs and upper arms, with the least representation in the cerebral cortex, not only evoke the least feeling in the amputee, but tend to have their symptoms disappear first.

One approach, the "peripheral theory," ascribes phantom limb sensation to irritants in the nerves of the stump. The "central theory" assumes that neighboring cortical areas stimulate the part of the brain, the sensory homunculus of the cerebral cortex, that once affected the phantom limb. Physicians agree that the brain can send physical sensations to the stump: Phantom limb sufferers are not imagining these sensations.

Psychological theories about phantom limbs tend to acknowledge the organic origins of the phenomenon, but they stress that amputation is a traumatic event for most individuals and that most patients are forced to redefine their self-concept after surgery. Many amputees, after surgery, feel that they are less than a whole person and feel anger and shame about their stump. Psychiatrist Thomas Szasz stresses the individual's need to preserve the "integrity of body image" and sees the phantom limb as both a form of denial and a means of recognizing the transition that must be faced by the recent amputee. The sensations are a way to focus attention upon the loss, but also to deny it ("If I feel it, it can't be missing").

Although the mere appearance of the phantom limb does not indicate any psychopathology, most psychologists feel that painful phantom limbs, not uncommon, tend to be a symptom of depression. Painful stumps are often symbols of anger, as well as grief, directed inwardly. Several psychologists have had much success healing painful phantom limbs by treating the patient as if he were "simply" a depressive.

Biofeedback has also worked in some settings, perhaps corroborating Szasz's theory that recovering patients need to focus on the stump in order to relieve anxieties about their self-image. Another clue to the supposition that problem phantoms might have at least a partly psychological base is the observation that many amputees forget about their phantoms for long periods of time but are immediately able to feel them when prompted by another person.

Why can't you ever buy fresh sardines in a fish market?

You are old enough. You deserve to know the shocking truth. There is no such fish as the sardine. The term *sardine* is actually a generic name for quite a number of different small fish. A fish doesn't become a sardine until it has been canned.

Different species are classified as sardines in various parts of the world. An international standard for canned sardines was developed by the Codex Alimentarius Commission. For practical purposes, the commission let every country establish its own definition of what constitutes a sardine and listed twenty-one different species as possible sardines. In Norway, sprats and immature herring are used for sardines; in South America, anchoveta is popular; in France and Portugal, young pilchards are the sardines of choice. The Codex standard was necessary, in part, because of squabbling among sardine producers. France and Portugal, in particular, maintained that sardines were not a generic product and that the sardine was a proprietary name reserved exclusively for the *Sardinia pilchardus*, the particular pilchard most popular in their countries.

The Bureau of Fisheries' position was that any fish in the clupeid family (small herring, brisling, sprats, and pilchards) may be packed and sold as sardines in the U.S. Anchovies may not be called sardines.

Sardines became popular in the United States after World War I, and this inexpensive lunch or snack was a sensation in the Depression era. But by the mid-1940s, sardine sales slipped and they've never regained their peak of popularity: At one time, the California sardine (the pilchard) was the largest fish crop in the country. The United States sardine industry has always been based in California and in Maine, which uses immature herring for sardines.

Since sardines are a cheap fish (indeed, before sardines caught on as a consumer item, they were used for fish meal), the main priority of the sardine industry is to catch a heck of a lot of fish and process them quickly. Let's look at the sordid death of a California sardine.

Sardines are caught in purse seines, huge nets with floats along the top edge and weights at the bottom. The seines are up to 2000 feet long, made of nylon and other rot-proof artificial fibers, and are often cast off of vessels 1000 feet long. Since the California sardine is phosphorescent, most fishing is done at night so that the sardines can be located by their movements near the surface of the water.

Once the sardines are brought to shore, the processing is almost entirely mechanized. Machines cut off the sardines' heads and tails in order to make them a uniform size. A vacuum then removes their viscera by suction. Another machine then automatically fills each can with sardines. At one time, the sardine industry loved to torture the consumer with soldered cans that could only be opened with an infernal key that seemed to be designed to open up wounds on the fingers rather than the can itself. Some of the imported sardines from Portugal, France, and Spain still come in soldered cans, but most other countries provide drawn cans that can be opened with a can opener. Many domestic sardines now come in "easy-open ring-pull cans," which actually are easy to open.

But the sardines' torture has just begun. After the can is filled, it is injected with live steam and heated up to a temperature that will expel trapped air in the can and coagulate and shrink fish protein. The purpose of this process is to eliminate much of the natural juice from the fish. By the time the can leaves this "exhaust box," it is approximately 150 degrees.

As the sardine continues along the assembly line, the can is tilted to remove the bodily fluids. At this point, seasonings are added (oils, brines, and sauces). Without these oils, sardines, unlike tuna, would not be palatable to most people.

The can is then sent to the automatic seaming machine for sealing. After the seaming machine, the can moves through a mechanical washer, which eliminates any pieces of fish or remnants of liquid that may cling to the exterior of the can. The can is then moved to a sterilizer.

The trimmings of the fish are not thrown away but are used for fish meal, fish oil, and condensed fish solubles. None of these products will ever reach a retail fish store, however. Although they are probably called sardines because the scientific name of some of the species includes the word *Sardinia, sardinops,* or *sardinella,* sardines have always been, at least in North America, a marketing concept rather than a particular fish. If it weren't for the can and the preservative capabilities of oil and salt, chances are most of us would never have heard of the sardine.

Why do we cry at happy endings?

Eureka! There is actually a conclusion upon which psychologists agree: There is no such thing as "tears of happiness." We cry not because we are happy but because unpleasant feelings are stirred up at the occasion of a happy ending.

Most adults are capable of repressing the urge to cry, but not without an exertion of psychic energy. When a happy ending indicates that our grief is no longer merited, the energy used to inhibit our tears is now discharged, sometimes in the form of laughter, but more often in an expression of the repressed sadness—tears.

Many people sit stoically through a tearjerker like *Camille* and then sob at a "heartwarming" thirty-second long-distance commercial or a reunion on *Truth or Consequences*. Happy endings often conjure up an idealized world of kindness and love that we once, as children, believed was possible to attain in our own lives. Children rarely cry at happy endings, because they are not yet disillusioned about their own possibilities.

For the adult, the happy ending is a temporary return to

the innocence of childhood—the tears stem from the recognition that one must return to the tougher "real" world. The child, without comprehension of the permanence of death, sees the happy ending as confirmation of the limitless possibilities of life.

The tendency to cry at happy endings is not restricted to stories. In real life, it is common for relatives of a critically ill patient to cry not before or during a delicate surgery, but after the operation is successful. The happy ending enables the loved one to feel safe in unleashing all of the sadness and anxiety that had been repressed.

Psychologists even dispute the idea that the tears shed at rites of passage such as weddings, graduations, and bar mitzvahs are tears of joy. Precisely because these ceremonies symbolize transitions in young people's lives, rituals stir up repressed anxieties in loved ones about the past ("Why wasn't my wedding as joyous?"), insecurities about the present ("Why haven't I found my true love like the bride and groom have?"), and fear about the future ("How will I survive when my children leave the nest?").

In our emotional world, we are needy, selfish, and demanding. We cry for *ourselves* at happy endings, not for others, but this does not mean we are incapable of feeling joy in others' happiness. Crying at the happy ending reveals our idealistic side, the part of us that yearns for the simplicity and love we once thought possible and the part of us that mourns its unattainability.

Why do White Castle hamburger patties have five holes in them?

In the competitive world of fast-food hamburgers, White Castle is the odd man out. It does everything "wrong" and yet it thrives. Here are just a few of the ways in which White Castle differs from the bigger players on the block.

- While McDonald's, Burger King, and Wendy's derive much of their income from franchising, all of the White Castles are owned by the parent corporation.
- While the big three have tried to spread as quickly and widely as possible through franchising, White Castle has been content to consolidate its operation in the Midwest, Kentucky, New Jersey, and New York. The first White Castle was built in 1921 in Wichita, Kansas, and the company is now based in Columbus, Ohio. Although White Castle predated McDonald's, there are more than ten times as many golden arches.
- While other fast-food emporiums have diversified their menus, offering breakfasts, gourmet sandwiches, and salad bars, White Castle has stuck with its staples. Yes, it offers a couple of other token sandwiches, but White Castle was even resistant to offering french fries and onion chips. Even now, hamburger sales constitute a full 60 percent of its gross, a much higher figure than any other major hamburger chain.
- While other fast-food chains emphasize the large size of their patties, mostly in an effort to lure more adults, White Castle sticks with its tiny burgers. When White Castle opened in the twenties, a Whitie cost five cents. Today, a burger costs around thirty cents. What you get for that thirty cents is a patty that is two inches wide and two inches long and not at all thick.
- Long before Wendy's, White Castle offered a square hamburger (and a square bun). The patties were not square for the marketing reasons that motivated Wendy's (see separate Imponderable) but for much more mundane reasons. White Castle's grills were designed to hold 30 hamburger patties. A square patty allows the burgers

to cook with literally no unused grill space, enabling the cook to increase turnover. Bob Goldberg, a spokesperson for White Castle, estimates that one store can produce 2500–3000 hamburgers in an hour, in the unlikely case it would be called upon to do so.

- White Castle burgers taste different than those of any of the other major chains and have a totally different texture. There is no delicate way of stating it: White Castle burgers reek of onions and have a soggy consistency. Far from an insult, these are the major reasons White Castle aficionados love them. Devotees call the little burgers "sliders," undoubtedly because one needn't chew a Whitie to ingest it; the bun and patty are soggy enough to slide down the digestive tract. The onion taste is unmistakable. Before the hamburger patties are to be cooked, a little water is placed on the grill, followed by onions. The beef is then placed over the onions and the buns over the beef. The buns are permeated by the rising, pungent steam of water, onion juice, and beef fat. Another reason why the buns turn out so soggy is that White Castle, which makes its own buns, deliberately makes them much lighter than other fast-food companies, so they can absorb the moisture more readily.

But of all the differences between White Castle and its competitors, none is as strange as one feature of every White Castle hamburger patty: It has five little holes. These holes were introduced in 1946, and as usual with White Castle, they serve a totally utilitarian function. The holes allow the steam and grease from the grill to escape up the holes to the upper bun, which cooks atop each patty. This release of steam and grease eliminates the need to flip over the meat in order to cook it evenly!

The holes help provide sliders with their texture. The rising grease helps give the upper bun the layer of gray film so beloved by customers, and the rising steam helps make the upper bun melt in your hand, as well as your mouth. The holes are not punched, but extruded by a specially designed instrument called a beef horn, which uses five steel rods to extrude the meat as the patties are formed.

White Castle has always been a family-owned operation, and like many private companies, the Ingram clan has been cautious in its policies. But the smallness and tight-knittedness of White Castle has also given it a strong sense of its identity. White Castle has not been ashamed of its customers, who are traditionally less affluent than the upscale types the Mc-Donald's commercials are aimed at. Many White Castles are located in inner city areas in midwestern cities that have supposedly seen better days, and many are open 24 hours and attract blue-collar workers on their way to or from graveyard shifts.

White Castle may seem like a nostalgic anachronism, the symbol of a bygone era. Don't bet on it. White Castle is doing very well, thank you. Its per-store gross in 1984 was $1.3 million, first in the fast-food industry (just edging out Mc-Donald's). Part of the Ingram family's conservative policy includes treating its employees well (at least by the philistine standards of the fast-food industry). While McDonald's, Wendy's, and Burger King have to contend with constant turnover of personnel, White Castle has phenomenal worker loyalty, partly because employees participate in a profit-sharing program.

White Castle doesn't think there are too many Imponderables about the hamburger business. It took a soggy, smelly burger that MBA types would probably deem unmarketable and created a mini-empire. Every decision it makes seems to have a practical, utilitarian purpose. If punching holes in its hamburger is going to increase its grill turnover, then holes in the hamburgers it will have.

Why aren't there seat belts in buses?

Buses, like all vehicles over ten thousand pounds, are not required by law to have seat belts. But why *don't* they? Although it is true that larger vehicles are better able to withstand crashes, the main reason most transit systems don't even try to install seat belts is that in all such experiments, passengers simply refuse to wear them.

Experiments with seat belts in school buses have been disastrous. Not only do children refuse to wear seat belts, but as an exasperated official of the National Highway Traffic Safety Administration told *Imponderables*: "In the hands of a teenager, a seat belt buckle is a lethal weapon."

After the first edition of *Imponderables*, we received a letter from Carol Fast, founder of the National Coalition for Seatbelts on School Buses. She pointed out that seat belts are now mandatory in over 150 U.S. school districts, one state (New York), and at least one big city (Denver), and that health care and consumer groups overwhelmingly support mandatory seat belts in school buses.

Fast also made the telling point that the main resistance to seat belts in school buses is an economic one. The school bus industry doesn't want seat belts because installing them will make their vehicles more expensive; the true scandal would be if this is the main reason why most school systems reject them.

The evidence of the efficacy of seat belts in preventing serious injury in buses is still inconclusive. Ms. Fast sent several studies that argued strongly for their use, but her organization suffered a setback when the National Transportation Safety Board, in March 1987, issued a finding that did not recommend requiring lap restraints on school buses. Among the NTSB conclusions: Twice as many students die each year getting on or off school buses than riding them; seat belts are unlikely to prevent deaths caused by head-on collisions (the cause of most on-the-road bus fatalities); money allocated for seat belts would be better spent on training bus drivers and providing better maintenance of the vehicles.

Why aren't there seat belts in taxicabs?

Actually, there usually *are* seat belts in taxis. If you dare reach under the seat cushion, you might find a seat belt among the debris. Taxis, like all automobiles, are required by federal law to be equipped with seat belts when they leave the factory. But most taxi commissions do not prohibit removal of seat belts.

Passengers hurriedly piling into back seats for short hauls find buckles more of a nuisance than a necessity. Even in localities that require seat belts in taxis, most cab drivers find that the path of least resistance is to hide seat belts behind the seat cushion—there is no law that seat belts must be easy to find. Many states have recently enacted laws requiring the *use* of seat belts by all occupants of cars, but taxis are usually excluded from such legislation and taxi jump seats always have been specifically exempted from needing seat belts.

Only when passengers demand and use seat belts will you find them in every taxi in the United States.

Why don't cats like to swim?

Many people think that cats are afraid of water. They're not. Occasionally, one can see a cat pounce spontaneously into the water.

Marlin Perkins fans can attest to the fact that many of cats' larger relatives, such as tigers and jaguars, love to swim. Jaguars are even known to dive into rivers and streams and attack alligators.

Abandoned house cats will dive into water to do a little fishing. So why isn't your cat likely to stick a paw into your backyard pool? For the same reasons your cat always drives you nuts: He has a cleanliness fetish, and he's lazy. Your cat, unlike your dog, refuses to have a good time and pay the piper. He won't get wet because he figures that it isn't worth the effort needed to dry and clean himself with his tongue to enjoy something as superficial as a marine frolic.

Unless you starve him and stock your pool with live herring, your cat is likely to remain landlocked.

Why does root beer taste flatter than colas?

The amount of carbonation in a soft drink is a crucial determinant in its taste, and individual beverage makers tend to be secretive about the exact amount of "gas" in their drinks. Still, there is remarkably little difference between different brands within the same genre of soft drinks.

Root beer tastes flatter than colas because it *is* flatter than colas. Carbonation of drinks is measured in gas volumes, the amount of gas a liquid will absorb at 68 degrees at atmospheric pressure. Most root beers measure about three gas volumes, meaning that the gas will occupy three times the amount of space in the bottle than the liquid will.

Here is the list of soft drink types, in descending order of effervescence, with their approximate gas volumes:

Type	*Gas Volumes*
ginger ale	4
lemon-lime	3.7–4
cola	3.5
root beer	3
fruit flavors	1.5–2

Without realizing it, many consumers might prefer fruit sodas primarily because of their relatively small carbonation levels.

If you dislike flat root beer, one suggestion will work wonders. Never pour a room temperature soft drink into a glass with ice—the radical temperature change traumatizes the poor gas—you will lose approximately half of all the carbonation. If you pour refrigerated soda into a chilled glass, you will lose only about 10 percent of the gas. Once the gas survives the initial burst of impact into your glass, it doesn't lose carbonation very quickly.

Although several friends have commented that some of the caffeine-free and sugar-free soft drinks seem less bubbly than their sugared and caffeinated counterparts, *Imponderables* was assured by all sources that this simply was not the case. None of the ingredients in drinks really affects the CO^2 content, since the gas is an additive and not a byproduct of flavorings.

Once in a while I hear bells going off in movie theaters. Am I going crazy?

Perhaps you are, but hearing bells in movie theaters is unlikely to be a symptom of impending mental illness. You are hearing warning bells that are going off to alert the projectionist that a reel change is imminent. Usually, the bell sounds two to two and a half minutes before the reel change is necessary.

You are less likely to hear these bells in newer theaters. Some theaters now use giant platters with continuous loops rather than smaller reels that must be manually changed. Some new projectors have simply eliminated the warning bells. Still other projectors are computer-driven, with controls that automatically trigger reel changes.

Whether or not your particular theater projection room has warning bells, in order to achieve a smooth transition during reel changes the projectionist probably relies much more on changeover cues that are contained in all standard release prints. Perhaps you have noticed these changeover cues, which consist of one black dot placed in the upper right-hand corner of the screen for four consecutive frames of the print. The first cue appears exactly twenty-two frames from the end of the reel.

Occasionally, the changeover cue is necessary when the

background of the screen is dark, making the cue nearly impossible to see. In this case, the black dot is usually surrounded by a thin white border.

The cue to actually start the second (idle) projector motor is the identical dot, placed in the same upper right-hand corner location, also for four consecutive frames. It is placed 12 feet and 6 frames, or about 8 seconds, from the end of the reel.

Two professional projectionists told *Imponderables* that after a few experiences with one movie, the changeovers become instinctive and that they don't even rely on the changeover cues at all. Both added that the warning bells shouldn't be audible to patrons and that if they were, the most likely explanation is that the projectionist has simply failed to close the door to the projection room.

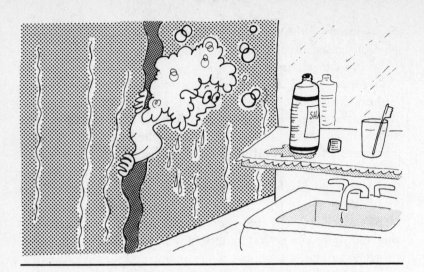

Why do some shampoos direct the user to apply them once? Why do some shampoos direct the user to repeat the application? Why do some shampoos direct the user to leave the shampoo on the hair for several minutes before rinsing?

Hair is dead. A good shampoo can make your hair more manageable, but it can't make your hair less dead. Any advertisement that claims a shampoo can nourish or revitalize your hair is, in effect, promising to feed and raise the dead.

The reason we need to wash hair is to eliminate dirt and grease. Every strand of hair is contained in a pore, the hair follicle, which extends well below the scalp. Every hair follicle also has a sebaceous gland, which manufactures an oil, sebum, to moisten the hair follicle, which in turn lubricates the hair and the skin on the surface of the scalp. Although oil is portrayed as the nasty villain on shampoo commercials, sebum prevents your skin and hair from having the dry and brittle consistency that plagues many elderly people. But sebum also collects bacteria and dirt.

The main purpose of shampoo is to eliminate excess oil and dirt that collect on the surface of the scalp. When shampoo hits the grease, it emulsifies the oil, turning fat globules into small droplets so that the oil and accompanying dirt will rinse out easily. No shampoo is capable of penetrating below the scalp. Shampoos merely collect and get rid of the material on the scalp and hair.

It is obviously in the interest of shampoo manufacturers to encourage buyers to shampoo every day, using two applications. A few decades ago, most people washed their hair every week or so, and a considerable number used their bath soap to do it. They didn't have many problems because excess oils don't accumulate that fast in most scalps. Although some companies recommend applying their product twice each time, the same shampoo will work admirably with only one application, for most people, if used on a daily or every-other-day basis.

Imponderables contacted several shampoo manufacturers. Each was quite willing to admit that there was nothing special about its shampoo that dictated applying it once or twice during the cleaning process and that the cleanliness of the hair was far more important than the properties of its shampoo in determining whether a second application was necessary. One company, a maker of "mild" shampoo, told *Imponderables* that it has repositioned its Neutrogena from an everyday product (with directions to apply the shampoo only once) to a non-everyday product ("rinse and repeat"). Notice that Neutrogena did not change the formula of its shampoo, but merely repositioned its marketing. Obviously, many will still use Neutrogena on a daily basis, but will now, as instructed, apply it twice rather than once. Neutrogena will probably sell more shampoo as a result. One shampoo, Ivory, from Procter & Gamble, specifies to repeat the application "if necessary." It isn't clear how someone in a misty shower is supposed to figure out if a second application is necessary, but Procter & Gamble cannot be accused of false labeling.

The nice thing about deciding which shampoo to use and how often to use it is that the decision is almost entirely inconsequential. Regardless of what your hairdresser tells you (or tries to sell you), no over-the-counter shampoo is going to wreck your hair. And none will make it beautiful. Even *Consumer Reports* can't get too worked up about differences among shampoos. In its September 1984, issue, *CR* had a panel rate 61 different shampoos. The result: no consistently strong preferences, and a pronounced tendency for women to like different shampoos than men. As an experiment, *Consumer Reports* added Octagon Dishwashing Liquid to its blind testing (dishwashing liquid can cost as much as ten times less than shampoos). The result: The dishwashing liquid fell a little below average in the panelists' preferences. *Consumer Reports*, as well as several marketers *Imponderables* spoke to, believes that the fragrance might be the most important ingredient in determining the success of a shampoo.

Some consumers are reticent about shampooing their hair twice a day and then using a conditioner. They notice some extra hairs gravitating toward their bathtub drain and wonder whether their shampoo and/or conditioner is to blame. A few more stray hairs may exit toward the drain, but shampooing doesn't do any permanent damage to your scalp. The average person sheds 50–100 hairs a day, and each of them will be replaced by the follicles. On any given day, 90 percent of your hair is growing (in what is called its anagen phase). The anagen phase lasts approximately three years. Some 10 percent of your hair is in its telogen or resting phase, which lasts approximately 100 days. In the resting phase, the hair follicle weakens and eventually the hair shaft comes out. Hormonal changes in the body are more likely to create shedding than is overshampooing. The reason our hair thins as we get older is that some follicles simply stop growing, not because we have abused our hair.

Scientists still cannot always pinpoint the exact cause, but many factors can create severe hair loss. In his book, *Everything You Need to Know to Have Great Looking Hair*, Louis Gignac lists the main causes of hair loss: "toxic amounts of Vitamin A; post-partum alopecia (hair loss after pregnancy, due to a shift in the hormonal balance); chemotherapy; stress and tension; certain hormone-altering drugs; thyroid imbalance; anemia; high fever; and diabetes." Local infections, some antibiotics, and cortisone can also precipitate hair loss.

Dandruff is a collection of scales of dead skin on the scalp. These dead skin cells, known as the keratin layer, peel off from other parts of the body as well and rise from the epidermis to the surface of the scalp so that newer, healthier cells can be regenerated. The keratin layer is most noticeable on the scalp because of the tendency of dandruff to shed on clothes and to contrast in color with dark-colored hair. Dandruff is a problem of the scalp, not of the hair, and as you might guess from its formal medical term, seborrheic dermatitis, is usually the product of overactive sebaceous glands rather than poor maintenance of the hair.

Many medicated shampoos, designed to help dandruff problems, direct shampooers to leave the lather on the hair for up to five minutes before rinsing. Although this instruction might seem like voodoo, there is a logical basis for the request. Dandruff shampoos contain active ingredients that need time to work. Salicylic acid, one of the most popular active ingredients in dandruff shampoos, promotes the peeling of dandruff. Zinc, another popular active ingredient (the active ingredient in Head and Shoulders), and sulfur also are effective against dandruff even though nobody knows why they work. For most people, regular shampooing with any product will help eliminate excess dandruff; although Head and Shoulders received higher than average marks from the *Consumer Reports* panel, many non-dandruff formulations fared as well or better.

When any shampoo eliminates dandruff, it does not eliminate the condition that caused the dandruff. Most dandruff shampoos work by having the active ingredient promote the peeling of flakes. By letting dandruff flakes fall freely into your bathtub drain, you keep them from falling all over your navy-blue blazer.

Why is June the most popular month for weddings?

Yes, June is the month when most schools break for vacation and when weather is appropriate for outdoor weddings, but these truths don't explain why June has been, since ancient times, a popular month for weddings. Like many contemporary customs, the popularity of June weddings has mythological origins.

The month of June is named after Juno, the Roman goddess of marriage and young people. Juno was reputed to take a proprietary interest in couples married during "her" month. An ancient Roman proverb counseled: "Prosperity to the man and happiness to the maid when married in June."

There is a more compelling reason why so many weddings are performed in June: May has been long considered the unluckiest month to marry. The sentiments expressed in the superstitious couplet "Marry in May, and rue the day" probably date back to Roman times. The month of May honors Maia, a Roman earth goddess, the consort of Vulcan and, most damaging to her month's wedding public relations, the patroness of old people.

Evidence indicates that the main reason June weddings proliferated in modern times was that the superstition about May marriages enabled June to fill its own quota plus some of May's postponed weddings as well.

By the twentieth century, the fear of May nuptials subsided considerably in the United States, but the glamour of June weddings remains. Juno has no reason to be complacent, however. For while June still attracts a disproportionate *quantity* of weddings, June brides and grooms have the same divorce rate as those poor unfortunates who marry in the month of Maia.

Why do other people hear our voices differently than we do?

We have probably all had this experience. We listen to a tape recording of ourselves talking with some friends. We insist that the tape doesn't sound at all like our voice, but everyone else's sounds reasonably accurate. "Au contraire," the friend retorts. "Yours sounds right, but *I* don't sound like that." According to speech therapist Dr. Mike D'Asaro, there is a universal pattern of rejection of one's own voice. Is there a medical explanation?

Yes. Speech begins at the larynx, where the vibration emanates. Part of the vibration is conducted through the air—this is what your friends (and the tape recorder) hear when you speak. Another part of the vibration is directed through the fluids and solids of the head. Our inner and middle ears are parts of caverns hollowed out by bone—the hardest bone of the skull. The inner ear contains fluid; the middle ear contains air; and the two are constantly pressing against each other. The larynx is also surrounded by soft tissue full of liquid. Sound transmits differently through the air than through solids and liquids, and this difference accounts for almost all of the tonal differences we hear when reacting negatively to our own voice on a tape recorder.

When we listen to our own voice while we speak, we are not hearing solely with our ears, but also through internal hearing, a mostly liquid transmission through a series of bodily organs. During an electric guitar solo, who hears the "real" sound? The audience, listening to amplified, distorted sound? The guitarist, hearing a combination of the distortion and the pre-distorted sound? Or would a tape recorder located inside the guitar itself hear the "real" music? The question is moot. There *are* three different sounds being made by the guitarist at any one time, and the principle is the same for the human voice. We can't say that either the tape recorder or the speaker hears the "right" voice, only that the voices are indeed different.

Dr. D'Asaro points out that we have an internal memory of our voice in our brain, and the memory is invariably richer than what we hear in a tape recorder playback. Although there seems to be no consistent pattern in whether folks hear their voices as lower or higher pitched than other listeners, there is no doubt that internal hearing is of much higher fidelity than external hearing. Listening to our own voice on a tape recorder is like listening to a favorite symphony on a bad transistor radio—the sound is recognizable but a pale imitation of the real thing.

Why do so many old people eat at cafeterias?

Enter almost any cafeteria in an urban area. Chances are you will see a disproportionately high number of elderly customers. The reasons for this phenomenon are many and date back almost a hundred years.

The cafeteria is an American invention. The first was opened in 1895 by Ernest Kimball, in Chicago. In 1899, he moved his cafeteria to the basement of the New York Life Building, where it was located until 1925. The period of the greatest growth of cafeterias was the 1920s and 1930s, noncoincidentally, perhaps, when today's seniors were young.

Cafeterias during the 1920s and 1930s were decidedly no-frills affairs. They were usually huge, with high ceilings, high noise levels, no decor, virtually no service, and huge menus. Customers were not encouraged to dally, and the unpleasant-ness of the ambience prodded customers to vacate their tables rapidly. The bare-bones wage scale paid by cafeteria owners ensured that whatever service provided was not old world elegance.

Cafeteria operators are faced with many challenges that did not beset their counterparts fifty years ago. There are many more options for low-cost eating now, and the particularly tough competition comes from fast-food outlets and employee cafe-terias in large factories and office buildings.

Yet there is now more excitement and growth in the cafe-teria industry than at any time since World War II. The new cafeteria chains are attracting a large share of young families, potential customers who might have been written off to Mc-Donald's in the past. Their most loyal patrons, however, are still older, retired persons. According to National Restaurant Association statistics, a person over sixty years old is 50 percent more likely to eat at a cafeteria than the population as a whole.

Cafeterias have suffered from a severe image problem. Downtown cafeterias changed little over the years; they were dinosaurs, remnants of a time when both rents and labor were cheap. If cafeterias were to survive, they would have to move where their middle-class customers moved, the suburbs, and near where they played, in malls and shopping centers. Most of the cafeterias that survived realized they had to market food as well as serve food, so they redecorated their dining rooms, carpeted floors, adorned their walls, upholstered their seating, and softened the lighting—ambience became a selling point for the cafeteria, for it appealed to people disenchanted with the manic and garish fast-food environments.

Not only have these facelifts worked, but there are several successful cafeteria chains, all based in the Sunbelt, that comprise one of the fastest growing segments of the restaurant industry. Such companies as Morrison's, Luby's, and Piccadilly's have prospered by keeping down their labor costs (one of the biggest reasons they are located in the South is the relative ease in using non-union help), and by placing themselves in shopping malls and free-standing suburban locations, avoiding outlandish inner city rents. Cafeterias now represent about 4 percent of all restaurant units in the U.S., but almost 6 percent of all commercial food and drink sales.

Throughout all of these changes in cafeterias, old people remained their most steadfast customers. Sure, some of their partisanship for cafeterias can be explained by a certain nostalgia for the type of restaurant that was a craze in their youth, but there are many more powerful reasons:

Demographics

1. Cafeterias are located where old people are located. Older cafeterias tend to be in downtown areas, where seniors are disproportionately represented. Newer cafeterias tend to be in middle-class suburbs and the South, where there is also a greater than average concentration of the elderly.

2. Next to baby boomers, the 65+ age group was the fastest rising between 1970 and 1980, and the trend promises to accelerate.

3. Contrary to popular belief, seniors dine out slightly more than the population at large—almost two times a week.

4. Although the distribution of wealth is wildly uneven among older people (with more than their share of the very rich and the very poor), the median income for seniors is rising faster than that of the population as a whole. Seniors have more discretionary money to spend, despite the problem of inflation for people with fixed incomes.

5. Retired people do not eat at cafeterias at work or, as young people often do, at school.

Price

1. A full meal can be assembled at a cafeteria for much less than most full-service restaurants and for not too much more than a fast-food establishment. In all surveys, seniors mentioned price as the number one reason they liked cafeterias. The average lunch ticket at a cafeteria is about $3, and dinner around $5. The eating preferences of seniors (see below) also help make a cafeteria a particular bargain. In most cafeterias, tipping is not necessary, another saving for the consumer over coffee shops and diners.

Eating Preferences

1. Seniors eat less food than any age group but small children. Many complain that portions are too large in full-service restaurants. Older people overwhelmingly prefer ordering à la carte for this reason. Morrison's has been particularly strong in recognizing this desire for flexibility in ordering. They offer a Savor Plate, which offers smaller portions of a full meal, with a choice of entree, two vegetables or a vegetable and potato, for one price. The cafeteria is the one type of restaurant that doesn't financially punish people with small appetites.

2. Older people generally prefer "home cooking" to "prefab" food. Many cafeterias cook virtually everything from scratch and even bake on-premise.

3. Seniors tend to stay longer than younger patrons. At cafeterias, there aren't sulking waiters intimidating customers into hurrying along. The slower, unpressured atmosphere at the cafeteria is a major part of the appeal of cafeterias to seniors. Although the turnover of tables is crucial to restaurateurs, older patrons compensate for staying longer by their predilection for eating earlier than young people. Old folks often eat dinner two hours before the dinner rush, and since most cafeterias stay

open continuously from breakfast or lunch through supper, they are occupying tables that might otherwise be vacant.

4. Cafeterias tend to have a larger selection of food than full-service restaurants in the same price range. Many cafeterias offer a hundred or more items every day. Restaurateurs told *Imponderables* that although not quite at Morris the Cat level, older customers tended to be more finicky about the quality and selection of their food.

5. Cafeterias feature foods that are only afterthoughts at other types of restaurants in the same price range. There are four foods that seniors order in restaurants at a much higher proportion than the population as a whole: coffee, vegetables, fish, and salad, in that order. All but coffee receive perfunctory treatment at most coffee shops but are staples at cafeterias. Seniors also order more dessert than average. Whereas younger people might go to a separate establishment to buy ice cream or pie, seniors tend to order dessert at the cafeteria, and this is one of the main reasons seniors tend to spend 5–10 cents more per capita at cafeterias. The one item that seniors seldom order is soft drinks—they buy only one-third as many as the average customer. The three popular food items that the elderly eat in the smallest proportion—(in order) pizza, hamburgers, and ice cream—are all foods that are not popular in cafeterias and that tend to be served by specialty restaurants catering to younger customers.

6. As a group, the elderly tend not to be adventurous eaters. One rarely encounters barbecued alligator at a cafeteria.

Psychological Factors

1. Many elderly people are widowed or alone for some reason. Some may not have a close network of friends. For people of any age, dining alone is not the most pleasant experience, particularly in full-service restaurants, where management and service personnel sometimes make single patrons seem like interlopers. To folks alienated from fast-food establishments, cafeterias are congenial places. Several sources indicated that

although older patrons tend to be fussy eaters, quality of food was not as important to them as the quality of the dining experience. For many people, eating out represents a vital and enriching form of social contact. There is something homelike about the personal contact a customer has with personnel who have actually prepared or served their food. Unrelated seniors can eat according to their individual dietary and economic desires. Single seniors find new friends at cafeterias without any pressure to do so. Patrons can commune or merely consume, depending on their preference.

2. While the stereotype is that old people are rigid, in fact old people are loyal. If a cafeteria, or any restaurant, does a good job, older customers tend to reward the establishment with steady patronage. Since many cafeterias have catered to older people for a long time, they have been rewarded with steady patronage.

If trailer parks didn't exist, would tornadoes exist?

Only if there were local TV newscasts to shoot pictures of them.

Where do they get that awful music for ice skating?

There are few things quite so disconcerting in sports as watching a pair of graceful and athletic skaters putting their hearts, souls, and bodies into interpreting music that sounds like common garden sludge. But, honest: Skaters and their coaches don't deliberately go out and select the worst arrangements of songs to showcase their prowess. There are logical explanations for why the music in free skating and ice dancing is often so unsatisfactory.

1. *The skaters must choose music that will please judges.* Competitive skaters are mostly in their teens and twenties, but judges are mostly middle-aged and older. Skating to "Stairway to Heaven" or John Cage's last avant-garde symphony is no more likely to score high with international skating judges than singing a Prince medley would with Miss America judges. In some cases, coaches select the music for skaters; some skaters select their own. Usually, it is a collaborative process between the two. But music must always be selected that will impress judges,

even if the arrangement is unlikely to bowl over audiences.

Theoretically, the selection of music should have no bearing on the marks that skaters are given. Judges are only supposed to judge how well skaters *interpret* the music. But not one of the skating authorities *Imponderables* spoke to doubted that music selection was a crucial element in building a successful program. One judge we spoke to admitted that an emotional, easy-to-follow piece of music, for example, was much more likely to engage judges than a difficult, abstract, cacophonous one. At the very least, pleasing music can keep the judges' attention during indifferent stretches of a skating program.

Skating judges need not have any musical training and are often musically parochial and unsophisticated. In international competition, it might be necessary to please (or at least not offend) judges from Japan, East Germany, and Canada. As a result, the television principle of "least objectionable program" is in force—if you don't want any of the judges to "tune out," you had better not offend any of them.

There have been several examples of prominent skaters being punished for their selection of music. The most notable recent example was the United States' ice dancing pair of Judy Blumberg and Michael Seibert, whose lovely performance of "Scheherazade" was not awarded a medal in the 1984 Winter Olympics because a judge felt that the music was inappropriate for ice dancing. According to Blumberg, this particular judge loved their skating, but didn't feel that any music inappropriate for a dance floor should be used in the ice rink.

2. *Vocals cannot be used in ice dancing and, in practice, aren't used in free skating.* Although there is no actual prohibition of vocal recordings in free skating, no one dares defy the common law. The vast majority of popular music contains vocals. The restrictions on vocal selection hinders skaters in two ways. The most obvious is that they are deprived of using what are often the best orchestrated and produced versions of any given song and the versions with which both audiences and judges are most familiar. Even worse, skaters and coaches are forced to become

musical archivists, desperately scrambling to find *any* all-instrumental rendition of a song they select. Judy Blumberg told *Imponderables* that when she and Michael Seibert were hunting for selections for their Fred Astaire-Ginger Rogers routine, there were some songs that they wanted to use for which they could find *no* all-instrumental versions.

The skaters and coaches we spoke to were evenly divided about the merits of the vocal restriction. Coach Ron Levington, who has guided the careers of Peter and Kitty Carruthers, feels that the rule isn't necessary and is rather arbitrary. In his opinion, the rule remains in force largely to distance competitive amateur skating events from the professional ice shows. Levington echoed the feeling of others that there is no reason that vocals have to be distracting to the judges.

Seibert and Blumberg's coach, Claire Dillie, strongly dissents. The purpose of ice dancing, to her, is to interpret rhythms. Dillie feels that there is a strong danger that if vocals were allowed in competitions, skaters would start interpreting lyrics rather than music and that skating doesn't need vocals and lyrics any more than ballet does. Still, Dillie recognizes the problems in finding suitable instrumental orchestrations of popular music, and she encouraged Blumberg and Seibert to use a piece that was specifically written and arranged for them.

3. *There is pressure to have several tempo changes in the long program.* The answer to this unstated law, which is meant to assure the athletic and interpretive versatility of champions, is for skaters to splice together up to four different pieces of music. Often, the surgical procedures help to showcase the performers but are musical abominations. There is a maximum of four-tempo changes per long program, but using four selections in as many minutes usually defies any attempt to make the routine an artistic whole. This is why "theme routines" like Blumberg and Seibert's Astaire-Rogers medley makes sense—they combine different songs and tempos but provide a sensible context for lumping together the different songs.

4. *It is important that musical selections provide opportunities for skaters to demonstrate difficult technical moves.* One reason ice-

dancing music is usually superior to free-skating music is that the latter competitions include more acrobatic leaps and spins. Music must be found to accommodate these movements, making the selection of appropriate material that much more difficult.

5. *Many arenas that house ice skating competitions have horrendous sound systems.* And when you are hearing the lousy sound secondhand via television (which itself has lousy sound), often with Dick Button wailing away, no music is likely to sound celestial.

6. *Too many skaters and coaches just don't care enough about musical selection.* Many skaters, with excellent technical skills, are just kids. They don't have the emotional depth to genuinely interpret music. Without a sensitive coach who understands their limitations and strengths, the students can become automatons, oblivious to the music. And if the coach doesn't care about music, the selections can be as banal and the arrangements as overbearing as many of them are today.

When we first asked experts about this question, we almost expected them to say that there was a special recording studio in West Germany that recorded muddy, ponderous, Euro-pop versions of songs specifically for skaters. Their strategy would be to drain the original song of all vestiges of life to ensure that the music would never distract judges from the skaters.

But it turns out not to be true. The skaters and their coaches really are trying to find the best music possible for their performances. But fate has conspired to make this a difficult task indeed.

Why are so many measuring spoons inaccurate?

The next time you strain to measure exactly one level quarter teaspoon of oregano for your stew, do yourself a favor. Relax. Chances are your measuring spoons aren't accurate anyway. Watch Julia Child throw spices around on television. She doesn't know from measuring cups.

For eons, home economists have told us that recipes in cookbooks refer to level measures. And for just as long, we amateur cooks have decided to throw a little extra in anyway. If two tablespoons of oregano is good in a spaghetti sauce, wouldn't two and a half be better? Or two heaping tablespoons (which can measure more than three tablespoons)? Occasionally, a stingy cook can actually be seen using level measurements for liquid, but no one can seem to resist using heaping measures of dry things. So it isn't like the inaccuracy of measuring spoons is a burning issue in America. Still, the nagging question remains: Why do kitchen measurement gadgets fail at the one and only task that they were created to do, the sole purpose of their existence?

If you have more than one set of measuring spoons or cups, try comparing them and see if they hold exactly the same amount of liquid. Chances are, they won't. The most obvious explanation for their failure to provide accuracy is shoddiness in craftsmanship and apathy from the manufacturer. This is probably true, especially since manufacturers have no federal testing standards to meet. A second hypothesis, equally valid, is that consumers never bother to test the accuracy of their measuring instruments. Also true. Do you even know who manufactured your cups and spoons? Who do you complain to? Woolworth's? A third possibility is that these instruments get altered and warped by contact with hot water in the dishwater. A possibility.

But notice something strange. Inaccurate measuring equipment almost always errs on the low side. *Imponderables* found measuring spoons that held as much as one-third less liquid than specified. Might this be the manufacturers' way of getting back at us cretins, who always use a heaping measure of anything we dump into a recipe?

Why don't we ever see the money from pay phones being collected?

A quick survey of twenty friends throughout the country revealed the startling information that a grand total of *one* had ever seen a pay phone being emptied. Do coins from pay phones roll down the slot and collect in underground passages? Are pay phones emptied at 4:30 A.M., when only cockroaches and Bianca Jagger are awake?

No such luck. This Imponderable has a rather mundane answer, one that doesn't totally explain this mystery. It turns out that almost all localities have their pay phones collected during normal business hours, from 9:00 A.M.–5:00 P.M., usually only on weekdays.

The operation itself is not a complicated one. There is a cash compartment on the right-hand side of pay phones. Each has a faceplate with a separate key, and the receptacle that holds the coins is removed. It should take the telephone company worker only about three minutes to empty a pay phone and leave, one reason why your chances of seeing a phone emptied, unless it is in a cluster of phones, is small.

Like jukeboxes that keep a tally of which records are played most often, the pay phone has a mechanism to keep track of usage patterns. The Coin Operation Information Network keeps a history of the last fourteen collections and predicts the optimum time to next empty the phone. The number of times a pay phone has to be emptied obviously depends upon the traffic pattern of its location—airport phones must be emptied much more often than one in a secluded street—but many phones need to be emptied as infrequently as every other week.

If anything, pay phones will have to be emptied less in the future. An increasing majority of pay phone revenue comes not from local calls but from credit card calls, collect calls, third party calls, and long distance calls through other carriers, all requiring no coin outlay from the patron.

Perhaps the reason we see so few phones emptied is that they tend to be isolated, whereas parking meters tend to be in rows—with phones, there is simply less of a chance of finding a worker unloading the apparatus. If you stood in front of one pay phone (that is emptied once every other week) for a minute, the odds are 6,720 to 1 against seeing a worker in any phase of emptying the phone. Stiff odds. Still, the sheer rarity of finding a pay phone being emptied seems a tad mysterious. We hung around street corners quite a bit in our youth, for more than a minute at a time. Where was the phone company?

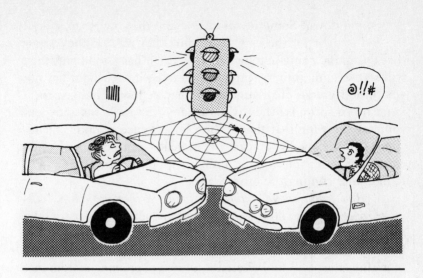

Why do yellow lights in Washington, D.C., traffic signals last longer than those in New York City?

Setting the timing of green, yellow, and red lights at traffic signals is a crucial job, not only does this decision regulate traffic flow, it can save lives. Miscalculation of traffic flow can lead engineers to set the wrong timing on traffic lights, causing at least four potentially serious problems:

1. *Unnecessary delay:* If the less busy street on an intersection is given the green light as long as the major road, the traffic on the busy street is likely to stack up with traffic. The more traffic on a street, the longer its green light should shine.

2. *Disobedience:* It might be assumed that the longer the yellow light and the longer the lag time between a red light switching to green, the slower (and the safer) the traffic will flow. This is simply not the case. Traffic-flow engineers have conclusively proved that drivers become attuned to the timing of traffic lights and will take advantage of any patterns that will

slow them down. Some drivers will crane their necks to look at the traffic light for the cars going the other way. If they know that the traffic going perpendicular to them has a red light, they will start edging into the intersection even if their light has not yet turned green. If drivers know that yellow lights last for a longish six seconds, they will run any yellow lights they encounter, knowing that they have a long "grace period." Disrespect for yellow lights can be a major cause of accidents, especially if the yellow light jumper happens to encounter a red light craner coming the other way.

3. *Bad traffic signal timing can induce drivers to use less adequate routes:* Traffic flow is planned to encourage drivers to take already busy streets, especially during peak hours. If drivers know that the lights have been set to create constant stop and go traffic, they will find alternate routes, overtaxing streets that were not meant for such extensive use.

4. *Accidents:* Particularly if the transition between green and yellow lights or yellow and red lights is too abrupt, rear-end collisions can result.

Many intersections are set up to maintain six different sets of timing for traffic signals:

- AM Peak Period Demand (morning rush hour);
- PM Peak Period Demand (evening rush hour);
- Average Daily Demand (in between the two rush hours and immediately after the PM Peak Period Demand);
- Nighttime Low Flow Demand (on major streets, it can be nearly impossible to hit a red light at night);
- Weekend Demand (depending upon the area, traffic might increase or decrease during the weekend);
- Special Function Demand (streets near ballparks, for example, would have radically different traffic patterns while a football game was in progress).

There is a seventh alternative. If traffic flow from opposite directions is greatly disproportionate some or all of the day, signals can be set to flash red in one direction and yellow in the other.

Even the number of flashes per minute is specified. A light should flash not less than 50 or more than 60 times per minute. According to the federal *Manual of Uniform Traffic Control Devices,* "The illuminated period of each flash shall be not less than half and not more than two-thirds of the total flash cycle."

How are all of these times determined? It isn't easy, but there are certain rules of thumb that engineers can use. For example, fifteen seconds of a green signal is generally recognized as the minimum time to accomplish major movement in an intersection. Thirty-second green lights are not uncommon.

Engineers must also consider the plight of pedestrians at an intersection, so pedestrian flow is measured as rigorously as automobile movement. The minimum period for a *walk* signal to flash, four seconds, is used when less than ten pedestrians are expected to cross (from both directions) on that green light. If ten to twenty pedestrians are expected to cross, seven seconds would usually be allowed, and longer if there are more pedestrians anticipated. Of course, pedestrians have a habit of stepping off the curb even when the DON'T WALK sign is flashing, the equivalent of the yellow light for cars. And just like a too long yellow light, if the DON'T WALK sign flashes forever, pedestrians will treat it like a green light, not only endangering their lives, but also slowing traffic wishing to make turns into the street they are crossing.

The situation is even more dangerous when determining the duration of yellow lights—clearly, the crucial timing decision on traffic signals. The first decision that must be made is how long the yellow light should be on. The usual range is three to six seconds. The longer interval is used when approach speeds coming up to the intersection are high, such as on undivided highways with traffic signals.

The next important decision is whether the yellow interval should be followed by a short all-way red interval, so that no new traffic will enter the intersection for a brief period. The purpose of having a four-way red signal is to clear out the intersection entirely before releasing cross traffic. The advantage of this strategy is that speeding cars will not encounter a just-turned green light and assume that they may blithely enter the intersection without looking for cross traffic. The disadvantage to using the four-way red strategy is that it slows down traffic and encourages the aforementioned red light craners to disrespect the meaning of the red light that they know will be turning green any second.

It is assumed that the reaction time of the average driver is one second. Thus, theoretically, if engineers allowed the signal for one direction to receive a green light at exactly the same time the opposite direction gets a red light, it should allow for a little more than a second to clear the intersection. Engineers in most urban areas, however, have found that with red light creepers, this strategy simply cuts it too close and have chosen a four-way red as the lesser danger.

The only way, then, to answer why Washington, D.C., yellow lights last longer than New York City lights (or any other major city that we have ever seen) is to note that D.C. drivers simply have a different psychology than New York drivers. Our capital's drivers will not enter and block an intersection even if they have a green light (it might be added that this behavior isn't all altruism—D.C. cops ticket drivers caught in the middle of an intersection, even if they had a green light). By eliminating gridlock, the D.C. police have created a situation where they can afford to allow drivers to "run" yellow lights. As long as the driver does not block the intersection, the yellow light can be "blown" without blocking perpendicular traffic when the light turns green. Ironically, then, the longer yellow light in D.C. promotes faster traffic flow than the shorter yellow light in New York.

Why does Wendy's have square hamburgers?

By the time that Wendy's International incorporated, in 1969, McDonald's was already a giant in the fast-food industry. Burger King was number two in the hamburger wars and was only beginning to market its product effectively. McDonald's had already secured so many desirable locations and could afford so much more national advertising than any of its competitors that Wendy's never seriously contemplated overtaking McDonald's in sales volume.

The Ohio-based Wendy's at first concentrated its efforts east of the Mississippi River and tried to find ways to differentiate its product from McDonald's. Right from the start, Wendy's signs read "Old-Fashioned Hamburgers." Many first-time customers must have been shocked to find that Wendy's considered an "old-fashioned" hamburger to be a square one.

Actually, Wendy's has a historical case. Although fast-food historians disagree, many credit Louie's Lunch, in New Haven, Connecticut, as serving the first hamburger in America. Louie's Lunch did (and does) make a square hamburger. Several of the United States' most venerable hamburger chains, such as White Castle and Little Tavern, also sell square hamburgers.

Wendy's square hamburger, however, was born far more out of marketing necessities than historical accuracy. Wendy's mission, if it was to be successful, was to differentiate its product from McDonald's. But how could it be heard when Wendy's couldn't possibly buy enough advertising time to drown out McDonald's messages?

One method to gain attention was to create a square hamburger. Clearly, a marketing advantage that Wendy's has always had over McDonald's is the relative size of its burgers. Wendy's strategy, from the start, was to focus the consumer's attention on the Wendy's patty. As Denny Lynch, vice-president of corporate communications, explains, "We think a square hamburger enhances the value perception when a customer sees the meat extending over the bun."

Of course, McDonald's also makes a quarter-pounder (which is Wendy's smallest sized burger), but Wendy's quarter-pounder *looks* bigger. McDonald's bun is larger in diameter and its patties are circular, both reasons why they don't re-create Wendy's slopover effect. Equally important, Wendy's cooking methods ensure that its hamburgers do not shrink as much as McDonald's. Every time a Wendy's hamburger is turned on the grill, the patty is pressed. The goal is for the hamburger to leave the grill with the same dimensions with which it was placed, raw, onto the grill. By pressing the hamburger at every turn, the patty might end up less thick, but it remains large in area. This pressing technique is much easier to perfect with a square hamburger than a circular one, where a circle could easily turn into an oblong or worse.

Another way in which Wendy's focuses consumers' attention on its patty is by using less condiment on its sandwiches. Even though Wendy's condiment selection is wider than McDonald's, if you order a quarter-pounder with "the works" at both places, you will notice the condiments more at McDonald's. Wendy's wants you, subliminally, to think that McDonald's has something to hide with its higher bun- and condiment-to-meat ratio. Of course, there is nothing inherently wrong with having proportionately more bread or catsup on a sandwich, except that customers often feel that they haven't gotten value from such a hamburger.

When Wendy's hit upon the "old-fashioned" slogan, it didn't refer to the squareness of the hamburgers as much as the fact that at Wendy's the customer always was supplied a custom-made sandwich, with choice of condiments, and a sandwich of a size more reminiscent of home than a fast-food restaurant. Although customers might not even consciously notice it, Wendy's felt it important to serve more costly onion rings rather than McDonald's diced or reconstituted onions. Wendy's offered mayonnaise as well as catsup and mustard, another sub-liminal signal that Wendy's is more generous, homey, and old-fashioned than McDonald's.

In the intensely competitive world of fast food, every decision is rationalized by extensive research. Once in a while, even in this day and age, somebody stumbles onto a successful strategy. The most important reason Wendy's hamburger is square, notwithstanding the discussion above, is that it was and is a terrific gimmick, one that differentiated Wendy's instantly from McDonald's and its clones. That the square hamburger was compatible with Wendy's eventually more sophisticated marketing efforts was a bonus, but most of all, the square hamburger got Wendy's noticed and created as much brand awareness and word of mouth as any advertising campaign. As Wendy's employees are fond of saying, "We don't believe in cutting corners!"

Why do most men part their hair on the left?

The most obvious answer to this Imponderable is that most men are right-handed, and it is easier for a right-handed person combing his hair to sweep his hand from left to right than vice versa.

The only problem with this solution is that most *women* are right-handed as well and a far smaller percentage of women part their hair on the left. We talked to several hair stylists, all of whom indicated that they were trained to place parts according to how the hair of the client naturally falls. Usually, this means that the hair is parted from the crown (the top part of the head) toward the lower regions of the scalp.

It would follow, then, that the crown of men's hair tends to be located on the left side of the head, except that *Imponderables* could not find any evidence to support this contention, either from cosmetologists or medical consultants. Have generations of fathers simply imposed their hair style on their sons? Do men have crowns more on the left side more frequently than women? As hair styles change, will there ever be a parting of the ways? The definitive answer remains elusive.

Why, despite the television and radio announcers' assurance that "for the next sixty seconds, we will be conducting a test of the emergency broadcasting system," does the test take less than sixty seconds?

The main reason is that nobody particularly gives a darn how long the EBS test takes. Almost all of the 9000 plus radio and television stations in the United States are voluntary members of the Emergency Broadcasting System, which is managed by the federal government. A few of the stations, with strong signals, are CPCS stations (Common Program Central Stations). In event of war or natural disaster, secondary stations monitor the CPCS station in their region in order to receive instructions about how to proceed in case of a disaster.

EBS stations are required to test that annoying tone once a week, and they do so faithfully. Actually, what you hear is the combination of two different frequency tones, joined together to assure that the sound is, let us be kind, noticeable. The choice and time of day of the test are random. The important thing to remember about EBS tests is that not much is being checked that you can't hear. The Federal Communications Commission and the Civil Defense Department are probably not monitoring the test. The EBS test is more like flipping on the air conditioner during the winter just to make sure it will work next summer.

Thus the stations are only concerned that they are able to create the desired two-tone frequency. If they are CPCS stations, they will have their own EBS generator to produce their tone; secondary stations, particularly radio stations, are likely to have prerecorded cartridges with an announcer's introduction and a silent tone. In any case, the duration of the tone itself is usually 20–30 seconds. The announcer's copy, which states that "For the next sixty seconds . . ." is lifted directly from a pamphlet issued by the federal government, but individual stations are not required to repeat it verbatim, and there is certainly no rule saying that the test must last exactly 60 seconds. The stations just know a compelling introduction when they hear one (and don't have to write their own), and have copied the sample text verbatim, even though they know the test takes more like 45 seconds than 60.

Why does chicken always take longer to cook than the recipe specifies?

The great chicken mystery was one of the toughest of Imponderables to unravel. Friends, with absolute unanimity, agreed that they invariably had to cook chicken longer than recipes suggested.

So we turned to some professional chefs, who asked if we took the chicken directly out of the refrigerator before washing it and putting it into the oven. "Yes," *Imponderables* replied. "Aha!" the chefs gloated. "Meat should be at room temperature or it will take longer than stated to cook."

We went back to the kitchen, let a whole chicken reach room temperature, and stuck it back in the oven. Now it cooked 33 percent slower than the recipe indicated, rather than 50 percent slower.

Undaunted, *Imponderables* contacted Perdue chicken with our problem and were stunned by their response. Perdue distributes millions of recipes with its packages of chickens, and although it receives many comments about its recipes, Public

Relations Coordinator Rita Morgan said, "I don't see a single reference to cooking time in recent years." Ms. Morgan added that specified cooking times were meant to be approximate and that they allow ten to fifteen minutes discretionary time to alert cooks to begin checking for doneness. The Perdue Oven Stuffer Roaster comes with a pop-up thermometer, like some turkeys, to eliminate the guesswork in timing. The pop-up thermometer eliminates one obvious problem—mistiming because of an inaccurate oven thermostat.

Could there be a difference between Perdue chickens and other chickens? Or was there a difference between Perdue chicken *recipes* and other recipes? *Imponderables* wrote to Holly Farms Poultry Industries, Inc., another giant among retail chicken suppliers in the United States, for some badly needed help.

William M. Rusch, director of marketing for Holly Farms, was kind enough to reply and, almost certainly, provide the solution to this Imponderable:

> We would suggest variation in chicken size to be a possible answer. More specifically, ten years ago, the average broiler/fryer weighed approximately two and one-half pounds. Today, the average is well over three pounds, with a range from three to four plus pounds.
>
> Older cookbooks and recipes invariably called for or the writer had in mind a two and one-half-pound bird. Still today we see recipes published calling for two and one-half pound chickens and would guess there are many republished recipes where no weight is stated, but the original assumption was a two and one-half pound chicken.

We went back to our cookbooks. Sure enough, they were all at least ten years old. Those books that did specify indicated a two and one half-pound broiler-fryer. Many recipes did not indicate the size of the bird.

When we cook a turkey, we use its weight to calculate cooking time, and yet many of us, evidently, think all chickens are created equal. As Mr. Rusch added, a smaller bird might cook faster, but the heavier the chicken, the proportionately more meat and less bone per bird, so that it would probably take longer to cook 1980s chicken parts than their scrawnier, earlier counterparts.

The "bigger bird theory" also explains why Perdue hasn't received complaints about its recipes. Its whole chickens range from about three pounds to over seven pounds for Oven Stuffer Roasters. But the recipes enclosed in packages are keyed to the size of that particular bird.

Imponderables conducted one final test. We cooked a six-pound Perdue Oven Stuffer Roaster (brought first to room temperature, of course) according to the recipe on the package. It worked. Done on time. One more Imponderable unplucked.

Why does unleaded gasoline cost more than leaded gas?

Knocking occurs inside the piston chamber during the spark of the piston cylinder when there is uneven combustion. The octane rating that differentiates "premium" from "regular" is nothing more than the measurement of the anti-knock capability of a gasoline.

Gasoline marketers had long added a tetraethyl lead (a.k.a. "lead") compound to otherwise finished gasoline in order to boost its octane rating. Why lead? It's cheap, and a small amount raises significantly the octane level of gasoline. The only problem with lead is that it polluted the environment. Hello, unleaded gasoline.

Use of unleaded gasoline should leave our air fresher, but at a cost. After the introduction of low-lead and unleaded gasoline, many customers were incensed that they were charged more for an ingredient being taken *out* of a product. This is one case, however, where the retailer actually is passing along higher costs to the consumer.

In order to retain the same octane rating, oil companies were required to supply a better quality gasoline. Without the cheap, artificial assist of lead, and without an economical alternative to lead at hand, the industry was forced to fight knock the old-fashioned way—with a purer, more expensive product.

Why are green olives packed in jars and ripe olives packed in cans?

For most food packaging needs, paper provides the cheapest protection against crushing. Paper is light, easing shipping costs, but offers little protection against the elements. For marketing purposes, paper remains one of the easiest and cheapest materials on which to directly print trade names and logos.

The tin can is the next cheapest package for mass production. Actually, 98.5 percent of a tin can is composed of sheet steel. There is only a thin coating of tin. Although the can better protects food from unsettling weather, it also covers up the product, requiring a printed label to reveal its contents.

The glass container, tin's main competitor, is heavier and more expensive, but it allows the consumer to see the food within. For certain corrosive liquids, glass containers are imperative. Olives are one food whose packaging has been determined more by health and safety reasons than marketing considerations.

Pick a fresh, unprocessed olive from a tree, pop it into your mouth, and you will be in for a most unpleasant taste sensation. Fresh olives are unpalatably bitter because of a glucoside that is neutralized by processing in the olives you buy in the stores. Most olives are grown for olive oil, with Spain and Italy together representing about one-half of the world market.

Spain dominates the sale of green olives in the United States for two reasons. Olive farming is Spain's largest industry: it is so important to the economy that olive production is heavily subsidized by the Spanish government. Spain has such an overwhelming head start in the marketing of green olives, with its huge acreage of olive trees and price subsidization, that American manufacturers can't compete in pricing. The other reason that the California olive farmers don't try to compete with Spain is that they can sell all the ripe olives that they harvest. Business is good—why try to take on the Spanish green-olive industry?

The same species of olives are used to make both green and ripe varieties, but the processing and packaging of the two are quite different. Green olives are picked just after they reach full size but before they have become fully ripe. At harvesting time, green olives have just begun to develop their color and haven't yet softened. As soon as possible after harvesting, fresh olives are placed in pickling vats. Olives contain a bitter alkaloid, oleuropein, which is removed by a lye solution. The lye solution is washed off the olives after permeating approximately two-thirds of the way toward the pit, thus allowing the olive to retain a hint of the distinctive bitter taste that fanciers enjoy.

After the lye treatment, the flesh of the olive becomes alkaline. The olives are then soaked in a salty brine. The fermentation process is slow. The processing of a green olive takes about two months.

Most imported green olives are packed by hand, which is why many brands feature fancy pyramid designs. The jars are then filled automatically with water or brine and then rinsed out automatically in order to eliminate sediment from forming on the inside of the glass. The jars are then refilled with brine and vacuum-sealed to retard growth of aerobic yeasts; some olive makers seal the jars at atmospheric pressure, usually without untoward results.

Molds are the enemies of all foods that are pickled and fermented. Molds metabolize the acid developed in the fermentation process. Olives can usually last for several months without spoilage; only if longer storage periods require more protection from the elements do olives need to be canned. Since consumers now refrigerate olives after opening vacuum-sealed jars, there is no technical reason why green olives can't be packed in jars.

Although glass jars are more expensive than tin cans, the Spanish olive industry has always favored them. Glass containers have less metal than tin cans to react with the food product itself, potentially affecting taste. Customers can see the olives decoratively arranged and, by sight, pick out pitted olives from those with pits, pimento from onion-stuffed. And although glass containers are more fragile than tin, they actually suffer less breakage in shipping than cans.

Farmers have always looked for ways to take the bitterness out of olives. In ancient days, olives were pickled in salt or treated with wood ashes to remove bitterness. Ripe olives are popular because they don't retain the bitterness of green ones. Although they are not pickled as long as green olives, their processing is just as important in determining the taste of the final product.

Black olives are not picked when they have reached their peak of ripeness; when they are, they tend to wilt when processed. In order to give the finished product some texture, ripe olives are picked when the fruit is anywhere from a cherry-red to a straw-yellow, depending upon the species. Unpickled olives are extremely sensitive to bruising, so freshly harvested ripe olives are stored in brine until the pickling vats are available.

As with green olives, the principal step in the pickling process is the elimination of the bitterness by exposure to lye solutions. The dilute sodium hydroxide used to process ripe olives not only makes the olives less bitter, but provides the fruit with its black color. If the lye applications are too intense or if the olives are canned at too high a pH value or washed at too low a pH value, the proper oxidation cannot take place and the desired color may not be attained.

Another danger in the processing of ripe olives is their tendency to shrivel while being cured in brine at the end of the pickling process. To avoid saturation, the olives are passed over a needle board to puncture the skins so that they will take the brine internally in measured doses.

At one time, brine was added by machine to fill cans already packed with ripe olives. Today, the cans are filled with extremely hot water and dry salt is added separately by an automatic dispenser.

As early as 1900, farmers discovered that ripe olives could be canned and preserved by heating after pickling. At first, they experimented with heating olives to 212 degrees. Most of the time, it worked. But there was one occasional but very serious problem—botulism. Heating to 212 degrees proved inadequate to kill all the bacteria. This same problem, botulism, is why ripe olives are found in cans today.

Ripe olives have a very low acid level—a pH value of 7.0. Low-acid foods, in general, are prone to botulism. In this case, the nasty contaminant is a mesophilic spore-forming anaerobic bacterium. To assure ridding the final product of the threat of botulism, high-temperature processing is a must. Canned olives are processed and sterilized, in the can, at 240–250 degrees for one hour.

Unlike early prototypes, current olive cans are lined with protective enamel. In plain tin cans, the color of the olives tended to bleach during storage—there is no problem with the enamel-tin can or with glass containers.

Yes, glass containers. The technology now exists to sterilize glass containers at high temperatures, but it is expensive, and there are only a few producers of ripe olives who do pack ripe olives in glass. The olives are sterilized in water in a retort with superimposed air pressure. With normal air pressure, the lids would blow off the jar.

California is responsible for the production of more than 90 percent of the ripe olives eaten in the United States (the rest are imported). Although technical and economical factors make tin cans more desirable for packaging of ripe olives, ripe olive makers feel they have lost little, since consumers don't find the dark, musky brine of ripe olives as eye-appealing as the transparent brine of green olives.

Why are typewriter keys in their current configuration?

The early developers of typewriters were obsessed with similarities between their keyboards and piano keyboards. Most typewriters therefore had eight to ten rows, since separate keys were needed for capital letters before the invention of the shift key. Most of these pioneer keyboards were arranged in strictly alphabetical order.

Christopher Latham Sholes, the American inventor responsible for the first production typewriter in 1873, found that the alphabetical arrangement of keys led to jamming when the typewriter bars of fast typists were on the upstrike. Sholes consulted his brother, a teacher, who developed the idea that the bars of letters used frequently in combination should come from opposite directions.

The brothers Sholes created the "QWERTY" keyboard we have today. The real purpose of this configuration was to avoid key jamming. At the time Sholes introduced the QWERTY keyboard, even the most proficient typists used two fingers—it was thought impossible to learn how to touch-type, even though the letters were arranged in alphabetical order!

Sholes, aware of consumer resistance toward new configurations of previously standardized products, sold his customers on the "scientific arrangement" of his new keyboard. Sholes claimed that the QWERTY keyboard required the least possible movements of the hands while typing. The exact opposite was the truth. Sholes's QWERTY arrangement necessitates a finger trek of great movement around the keyboard to form the most basic English words. Yet Sholes's misleading advertising is still believed by most typists.

With the advent of sophisticated typewriters, word processors, and computers that work without traditional bars, the need for the QWERTY configuration is gone. Many theorists have proposed more efficient letter arrangements, and the Dvorak keyboard has gained rabid adherents, but the question remains whether a century of QWERTY keyboard use can be overcome by such a mild force as logic.

Why does an *X* stand for a kiss?

Those cute little XXX's we affix to Valentine's Day cards and mash notes, with or without their companion OOO's, began not as symbols of affection, but as substitutes for signatures in the Middle Ages, when the vast majority of citizens were illiterate.

But the *X* was also used by well-educated people, who were quite capable of signing their names, and was found on even the most formal and important documents—wills, contracts, deeds, and proclamations. Even kings and queens signed with the *X* as a symbol of good faith—an oath that the contents of a document were true. In some cultures, an *X* became a compulsory binding oath—without it, a contract or agreement was considered invalid and not legally binding.

It was not an accident that the *X* was chosen as the substitute for a signature, and contrary to popular belief, it did not gain acceptance because of its simplicity for the illiterate.

The acceptance of the *X* had everything to do with Christian symbolism. The X was the sign of St. Andrew, one of the twelve apostles: signing the X implied a guarantee to live up to one's promises in that saint's name.

The X also had intimate associations with Christ himself. The X was regarded as a visual representation of the Cross of the Calvary and the Crucifixion, and X, as well as an English letter, is the first letter of the Greek word for Christ, *Christos*. (The Greek letter, of course, is chi.)

How did this legal and religious symbol metamorphose into a romantic one? To further guarantee the sincerity of intentions, people in the Middle Ages solemnly kissed their signatures, much as we put our hand on a Bible to swear to our veracity in court. This kiss became known as the "kiss of truth," and because the kiss finalized and bound many agreements, it spawned another saying that many think had romantic origins— "sealed with a kiss."

Over the years, as notaries public, literacy, and lie detectors lessened the need for the mark, the X lost its sacred connotations. It reached its peak in popularity in the early and mid-twentieth century. During World War II, the British and American military were so alarmed by its constant use that they forbade their soldiers from putting XXX's in their letters home, fearing that spies might insert cryptic codes into these humble marks, which once stood for truth.

Why is saffron ridiculously expensive?

The saffron threads used to color and flavor many dishes, particularly in Indian cooking, are the golden orange stigmata of the autumn crocus, a plant of the iris family. Autumn crocuses are far from rare. So why is saffron so dear?

There are two reasons. The crocus flowers must be picked by hand to extract the saffron threads. As many as 500,000 flowers

(1.5 million stigmata) are needed to collect one measly pound of saffron.

The flowers are picked immediately after they blossom, and the stigmata are cut with fingernails and then dried by the sun or by fire. During this drying process, the saffron loses approximately 80 percent of its weight.

Saffron could be cultivated in North America and still is grown in parts of the Mediterranean, but where could the U.S. or other affluent countries find labor inexpensive enough to produce saffron as cheaply as the "ridiculously expensive" price we pay today?

Why are there so few women pilots on commercial airlines?

Have you ever entered a commercial aircraft, fastened your seat belt, and heard a female announce, "Hello, this is your captain speaking"? Chances are, you haven't.

There are very few women captains piloting commercial planes, and the situation is not likely to alter much in the near future, even though there are more women who are first or second officers today. The demand for pilot jobs is immense, but there is a finite number of opportunities.

The founders and present-day executives of most of the major airlines have military backgrounds. Many of them saw and see their companies as paramilitary organizations. Traditionally, the talent pool for commercial airplane employment has been the military. The airlines were able to pick the cream of the crop from air force, navy, and marine ranks, pilots with thousands of hours of pilot-in-command time and experience with aircraft more sophisticated than they will probably ever be asked to fly as civilians. The academic curriculum of the military also tends to be more comprehensive than civilian education. Given an "all else being equal" alternative, how could the airlines be criticized for selecting the right stuff when the competition is mostly (male and female) flight instructors and corporate pilots?

Ever since the WASPS of World War II, women have flown transport planes for the military, but they have not seen combat action. There is no pressure on the military to change this policy. As a result, women (and nonmilitary men) are usually forced to pay for their own training, and to receive flight education that might be perfectly adequate, but less wide ranging and glamorous than their military counterparts.

WASPS from World War II did not break into commercial piloting. It wasn't until the late 1960s that a woman attained the rank of captain at a large commercial airline, when Emily Howell broke the ranks at Frontier Airlines.

Obviously, cultural gender stereotypes are an important reason why there aren't more women pilots. There are many other physical, hands-on technical jobs that have not attracted women. Women haven't traditionally flocked into the ranks of civil engineers, for example. But when affirmative action legislation was passed, the airlines were on shaky grounds on their hiring history of all minorities, and greater attention was given to all minority applicants. With the opening of doors to nontraditional candidates, the mix of military- to civilian-trained pilots also changed, so that now about half of all commercial airline pilots are trained outside of the military. Although it may be harder, financially and psychologically, for women to achieve the same level of piloting training as men, qualified women, if anything, have an advantage over men. The airlines look for qualified women—if not for altruistic reasons (they could face the prospect of government and class action discrimination suits), then for practical ones.

Some sources, off the record, indicated that the airlines think that customers feel insecure with women pilots. Passengers, so the theory goes, feel queasy when they hear a female voice announcing she is commanding the plane—some out of sheer sexism, others because they assume that no female could have enough experience to warrant having *their* lives in her hands.

In *The Right Stuff*, Tom Wolfe talks about how pilots everywhere emulated test pilot Chuck Yeager's verbal bedside manner. Pilots still imitate him. If our plane plummets thousands of feet, we can expect a husky male voice to drawl nonchalantly, "Ladies and gentlemen, we are experiencing mild turbulence. I'd recommend you keep your seat belts fastened until we correct this inconvenience. Isn't that a lovely view of Lake George on the right?"

This is our image of pilots. They are daring and macho, but also comforting and endlessly secure and confident. The notion of women pilots clashes with all sorts of cultural stereotypes. With far fewer opportunities to receive comparable training and little incentive or pressure on airlines to change hiring practices, the prospect for rapid change is small. Airline companies may now encourage men to hand out complimentary macadamias in the aisles, but women aren't particularly wanted in the cockpit.

Why do some Baskin-Robbins stores charge 5 cents more for an ice cream cup than for an ice cream cone?

Baskin-Robbins stores are all franchises. Although the B-R headquarters provides guidelines, it is up to each storeowner to set the price structure for his or her establishment.

Many years ago, it was common for Baskin-Robbins and other ice cream emporiums to charge five cents more for a cup than for a cone. At a time when a single scoop might cost twenty cents, a nickel toll for a cup and a plastic spoon seemed like an excessive penalty. Now that a single scoop at many stores costs a dollar or more, the nickel surcharge is still around, but not nearly as prevalent.

Could the cup and plastic spoon actually cost the store owner more than a sugar cone? The answer: yes. Ten years ago, the cup and spoon (mainly the cup) cost a couple of cents more than the sugar cone (which costs considerably more than the cup cone favored by kids who don't know any better). The B-R franchisees created the five cents' surcharge to cover their extra cost for the "packaging" of the cup. Inflation in the cost of cones has greatly exceeded the rate for plastic and paper products of late, and though your Baskin-Robbins dealer still pays a little more for cups than for cones, the difference is negligible.

Baskin-Robbins would prefer its franchisees not to charge extra for cups. The two-tier price structure might have had some justification years ago, but doesn't now, and consumers tend to feel they are, literally, being nickeled (and dimed) when they have to pay five cents extra for a cup when they have already laid out an investment of a dollar or so for one scoop.

An informal survey of several Baskin-Robbins franchisees, none of which now charges extra for cups, revealed that the pricing of their ice cream is based upon what their market can bear. There are regions of the country where customers would balk at not being able to get two scoops of ice cream and change for a one-dollar bill. Said one dealer, who stopped his cup surcharge five years ago, "I could no longer get away with it."

Many people believe that the reason for the higher cost of cups is that the serving of ice cream is larger. At Baskin-Robbins, at least, that is not the case. All scoops are supposed to be two and a half ounces, but generous dippers are the bane of ice cream profit ledgers. One franchisee estimated that he loses $25,000 a year because employees lay too heavily on their scoopers. Some ice cream dippers do tend to give bigger servings in cups, probably because a scoop is wider in diameter than a cone, but doesn't come close to filling the cup (this is undoubtedly why Häagen-Dazs cups are tiny—they feel that there is a marketing advantage to overfilling a small container rather than partially occupying a larger one).

One reason customers might tolerate a higher price is that cups are particularly popular with two age brackets: adults and little children (whom experienced parents are too savvy to entrust with a cone and any edible substance). In both cases, the purchase is generally made by over-25 adults, who might be less price resistant than kids, teenagers, or young adults. Cups also sell best during the colder months of the year, when serious eaters aren't deterred from purchases merely because the streets are as frozen as the ice cream, let alone by an extra five cents.

Baskin-Robbins does not conduct system-wide research on this topic, but our poll indicates that almost half of all scoops that Baskin-Robbins serves are in cups, so that the loss of the potential five cents' gross on the cup surcharge is significant and must be absorbed by raising the price of the other products. The cup surcharge was never really a consumer rip-off. If anything, B-R franchises were fairly passing along extra costs directly to the people who used the "frill." And now that cups don't cost the retailer more, they have stopped a practice that made some customers perceive them as cheap or usurious.

For another food, the cup surcharge is back. Many frozen-yogurt stands charge more for a cup than a cone. They, too, have justification for the practice. They put more yogurt into a cup than a cone. To anyone who has ever attempted to eat a large frozen-yogurt cone and keep his clothes and body stain-free, the reason yogurt cones tend to be small is no Imponderable.

How do they determine on which corners of intersections to put street-name signs?

The actual placement of street signs is usually left to local governments, but the guidelines laid out in the federal government's *Manual of Uniform Traffic Control Devices* is almost always respected. This manual is amazingly specific. It specifies, for example, that the lettering of street-sign names should be at least four inches high. Supplementary lettering, such as "St." or "Ave.," must be at least two inches high. It commands not only that all street signs should be either reflectorized or illuminated, but that the legend and background of the sign should be of contrasting colors and that it should have a white message and border on a green background.

The manual recommends that all intersections should be equipped with street-name signs. In residential areas, the minimum is one street-name sign per intersection, but no guidelines are stated for which corners to place the signs on.

For business districts and principal arterials, however, the manual is much more precise and answers our Imponderable specifically:

> Street name signs should be placed at least on diagonally opposite corners so that they will be on the far right-hand side of the intersection for traffic on the major street. Signs naming both streets should be erected at each, located and mounted with their faces parallel to the streets they name.

If a busy intersection has signs on all four corners, it is probably the judgment of a local engineer that the traffic load justifies the added expense.

Why does Roger Ebert receive top billing over Gene Siskel on *At the Movies*?
Why did Gene Siskel originally have top billing over Roger Ebert on *At the Movies*?
Why did Roger Ebert receive top billing over Gene Siskel on *Sneak Previews*?

Billing in motion pictures and television has become such a crucial issue for performers and their agents that all sorts of devices have been invented to avoid the ultimate confrontation: Who will receive top billing?

With a large cast, the most common tactic to indicate equality of billing is to list the cast in alphabetical order. Alphabetical order suggests an ensemble troupe bereft of ego or petty jealousies. Many actors who wouldn't dream of accepting second billing in a movie will gladly lose themselves deep into the alphabetically ordered credits of an off-Broadway dramatic production.

It is more of a problem to signify equality between costars. No production yet has stooped low enough to list two stars with the caveat, "in alphabetical order." But one solution has been devised to display equality graphically—the left-down/right-up configuration, currently used on *Cheers:*

Shelley Long

Ted Danson

Since we read from top to bottom and left to right, our eyes presumably become gridlocked directly between the two stars. One wonders how they decided which actor's name would go where if there was no distinction in status between the two positions. (Note to Ted Danson: An unscientific survey indicates that most people see Shelley Long's name first when reading the credits. Call your agent.)

Roger Ebert and Gene Siskel are both respected film critics. On their PBS series *Sneak Previews,* Roger Ebert received top billing over Gene Siskel. When Ebert and Siskel left PBS for the greener pastures of syndication with *At the Movies,* Siskel assumed top billing. Was there a bloodless coup? Did Siskel triumph after a protracted, bitter contractual wrangle?

When the fall 1984 season of *At the Movies* began, Siskel's name displaced Ebert's as top-billed. Had Siskel counterattacked?

No such juicy story. Roger Ebert was kind enough to answer this Imponderable personally:

I was top-billed on "Sneak Previews." For "At the Movies," we flipped [a coin]. He won; he gets top billing the first two seasons. Starting September 1984, it's Ebert and Siskel for the next two.

Ratings willing, they'll flip (a coin or billing) in September 1986.

Why are there two title pages in most books?

Open almost any book, including this one, and you will find that the first printed matter is likely to be the title of the book printed simply on the right-hand side. Turn the page, and you are likely to find the "official" title page, with the name of the author included, and the name of the publisher, or its imprint, at the foot of the page.

What is the purpose of that first title page? Why start a book with such a drab opening when the title information and more is provided by the elaborate and larger sized "real" title page?

The name of the first title page is the bastard title or half-title. Most of the time, it doesn't serve any purpose. But it once did.

In the early days of publishing, before mass distribution, scribes hand-printed books and there was no such thing as a title page. Books were produced by commission, so customers, usually noblemen, already knew the title and contents of the book they were buying. Books then were bound immediately upon completion and were thus protected from fingerprints and dirt.

When books began to be printed rather than handwritten, booksellers were usually the publishers as well. Printers sent bundles of hand-tied finished copies to bookstores to be bound, leaving naked pages exposed to the elements. The top page of the book was most often damaged, just as the front pages of newspapers today sometimes get pulverized or soiled. In anticipation of the problem, printers started leaving the first page of each manuscript blank. The binder would then eliminate this page when the book was ready to be bound.

The only problem with leaving the top page blank was that it hid the identity of the book. When potential customers looked at books (which were usually custom bound to the specifications of the retail buyer), they turned over the blank page to see what the title was, again exposing the first page of the book to abuse.

To solve this ridiculous problem, printers started identifying the title of the book on the front page in a style resembling our current bastard title. Although this page was still intended to be cut before the book was bound, the same problem reoccurred—once the title page became a popular feature, customers wanted it in pristine shape; there was now a need for a sheet of paper to protect the title page, but also to identify the contents of the book. Thus the bastard title was born. The reason the half-title is so simple compared to the full-title page is that it was originally not even supposed to remain in the finished book. To indicate the less than noble origins of the half-title, it is called the *Schmutztitel*, or "dirt title," in Germany.

When printers began binding books immediately upon their completion, there was no longer a real function for the bastard title. The hard cover now protected the title page. The title page as we know it today, with the title, author, and publisher's name, became commonplace by the sixteenth century. The original purpose of the bastard title was probably forgotten, but it became and remains a traditional feature of most books.

Publishers, like most business folk today, are cost-conscious, and more books of late have been produced without bastard titles. Many publishers bind books only in increments of 8 or 16 finished pages. Many romance-novel series, for example, are always exactly the same number of pages. If a manuscript runs 250 words over the desired length, the book would have to be, not 1 page, but 8 or 16 pages longer than desired, thus substantially increasing the cost to the publisher. The type size could be reduced to make the extra words fit, but most likely those words will be cut. The only alternative is to fiddle around with what publishers call the oddments, all the stuff in the book that is not contained in the chapters. One of the most dispensable oddments is the bastard title, a feature that cover art and book jacket titles have long rendered obsolete.

Why is the right-hand side of a book always odd-numbered?

It is not difficult to trace our method of pagination. When books were first produced, there was no prefatory material whatsoever: no table of contents, no foreword, not even a title page. The first few paragraphs of the book served the function of introducing the subject of the text and its author. Since, when bound, the first page of a book is on the right-hand side, it is logical that page one would always appear on the right-hand side.

Logical, but not quite accurate. The earliest books were not paginated at all. The title page was not introduced until around 1500. Folios, or page numbers, did not appear until well after that.

As printing spread throughout Europe, introductory material in books became more and more elaborate, and rules were developed about the precise placement of this "front matter." From the beginning, the right-hand side was always used to display important information. Although readers may not be consciously aware of their placement in a book, we know through experience that all of the vital information we need before starting a book will be contained on the right-hand side. Here is the running order of the front matter as prescribed by *Words Into Type*, the premier reference book on such matters. Although not every book will contain each of these elements, they will be contained in this order and on the same side of the book. Included is the Roman numeral pagination usually included in the front matter. Notice that not only do the right-hand pages get to carry all the good stuff, but that left-hand pages will be left blank rather than be assigned the honor of possessing important front matter.

i. Right-hand page—the "Bastard Title," also known as the "Half-Title" (see above Imponderable)—usually the first printed matter inside a book. Generally discloses nothing but the title of the book.

ii. Left-hand page—the "Announcement" page—might include other books by the author or other books in the same series. But this page is usually left blank. Although almost obsolete, the frontispiece would be inserted here, if anywhere. Although the frontispiece is the one eye-arresting piece of front matter printed on a left-hand page, its function is to enhance the right-hand title page, which it faces.

iii. Right-hand page—Title page.

iv. Left-hand page—Copyright page. Yawn. When trying to save space, some publishers include the author's dedication here.

v. Right-hand page—Dedication.

vi. Left-hand page—Blank.

vii. Right-hand page—Table of Contents (some publishers place the Table of Contents and List of Illustrations after all of the other front matter).

viii. Left-hand page—Table of Contents (continued), List of Illustrations (if any), or blank.

ix. Right-hand page—List of Illustrations (if Table of Contents ran two pages) or Acknowledgments (if any).

x. Left-hand page—List of Illustrations (continued) or Acknowledgments (continued), or blank.

xi. Right-hand page—Editor's Preface (if any).

xii. Left-hand page—Editor's Preface (continued) or blank.

xiii. Right-hand page—Author's Preface (if any).

Immediately before the text itself begins, it is customary to place another half-title on the right-hand side, and if the author desires, an epigraph on the left side facing the first page of text. If there is no epigraph, the left-hand page (behind the half-title and facing the first page of text) remains blank.

All sections of the text start on the right-hand side. Many designers insist on starting all chapter headings on the right side—this is one design element that adds a patina of class to a book without the reader necessarily noticing why.

With the possible exception of the List of Illustrations, not one item in the front matter that a reader is likely to want to examine begins on a left-hand page (and some publishers always start even the list of Illustrations on the right-hand side). If the author's preface, for example, runs only one page, the next left-hand page will be left blank in preference to starting the foreword there.

Generally, all pages that are of the same paper as the text are included in the book's pagination. The pagination in most books disregards the Roman numerals in the front matter, starting the text with page one; a few begin in Arabic numbers from where they left off at the end of the front matter. The first page of text in many books is printed as page three, counting the half-title as the beginning of the book. Blank pages are always included in the pagination.

Many bibliophiles would rather return to the Middle Ages and eliminate page numbers altogether. These idealistic folks feel that because page numbers are set off by themselves in the margin of a page, they distract the reader's eye from the body of the text and pose as a nuisance to the peripheral vision.

But the battle has been lost, for readers clearly seem to like folios, which, combined with the index and table of contents, make locating desired material easy.

Clearly, all of these commandments to place important features of the book on the right-hand side indicate that there are powerful psychological and design principles at work. Newspaper designers are aware that the third page of a newspaper draws the reader's attention more than the second page. In most papers, page three is the "second front page," with page two containing the index and lighter features. The first right-hand page inside a magazine is prime advertising space, almost as valuable as a back cover, since most readers turn to it rather than the inside front cover first.

Imponderables could find no definitive explanation for why, in a culture that reads from left to right, the right-hand side of a book steals our attention. Besides the less than startling observation that we have become accustomed to the practice, the best answer might be the most obvious. The first page of a book (not including the cover) starts on the right-hand side. In order to find the first left-hand page, we must turn over the leaf and look at the back of the first page. Is the back of anything as important or eye-catching as its front? (Remember, we are talking about inanimate objects here.) When you hear the cliché "you can't judge a book by its cover," don't you assume the reference is to the front cover?

Why are there so many irregular sheets? And why are so many fancy-schmancy department stores willing to stock them?

If Dole has a pineapple that is wholesome but peculiarly shaped or not quite the desired color, it can use the fruit for juice rather than canning it as is. No branded product wants to alienate loyal customers or scare away first-time customers by providing an inferior product.

Linen makers are no different. Almost 10 percent of all sheets produced are defective in some way, usually a minor error in hemming, sewing, or printing. Jim Andes, a spokesperson for Cannon Mills, told *Imponderables* that his company has workers looking for imperfections throughout the production process. Cannon has managed to reduce its rate of irregulars to under 8 percent and doesn't expect that it can bring the rate down much.

What *is* different about irregulars in sheets and pillowcases, as opposed to just about any other product, is that consumers love to buy them and even the ritziest department stores are more than willing to sell them. And stores don't hide irregulars in the bargain basement. Two of the trendiest and most successful department stores in New York, Macy's and Bloomingdale's, featured irregulars on the covers of their latest advertising supplement in the Sunday *New York Times*. Why will consumers who would frown at buying "seconds" of shoes or blenders form a line to buy irregular linens?

Buyers have learned that irregulars do not have defects that affect the quality or the wear of the sheets. Although some customers think that irregulars are odd-sized, this is rarely the case. Most problems in hemming and sewing are undetectable to nonprofessionals. The sheet producers take out their labels so they won't be identified with the flawed merchandise and so larcenous types won't try to get full-price refunds for items bought at irregular prices, since most manufacturers will ordinarily unconditionally guarantee their sheets. Through sales of irregulars, the manufacturer gets a nice chunk of money from the retailer to help defray the cost of labor and materials. Although the manufacturer knows it will end up with some irregulars after any substantial production run, they are rarely sold to retailers in advance. More often, a retailer will inquire about the availability of irregulars while ordering first-quality merchandise. The linen houses can sell as many irregulars as they have with ease—irregulars are in great demand.

The department store might price a sheet that normally sells for $15.99 at $7.99 as an irregular. Obviously, their profit is less, so why do retailers bother? Because they've found out that cheap sheets are a tremendous traffic builder. Just as grocery stores have found it wise to keep the price of certain staples like milk and hamburger low (or risk being undercut by the supermarket down the block), so have department stores found irregular sheets (and to a lesser extent, imperfect towels) an inexpensive way to generate more foot traffic in the store. They are loss leaders of sorts, though few stores actually lose money on irregulars—they sell for less, but they also cost the stores less. Best of all, since consumers have come to think of themselves as crafty rather than cheap for buying irregulars, even classy department stores find that irregular-sheet sales increase business without tarnishing their quality image.

The linen industry has mixed feelings about irregulars. On the one hand, the sale of these items—that otherwise would not be sold—directly improves its bottom line and allows it to keep its prices low. On the other hand, sheet makers don't want the public to think of their product as a staple like milk or a light bulb—something that merely needs to be replaced when it is used up or worn out. Their greatest desire is for us to think of sheets as a "want" item, like a record album or a blouse—something we actively covet rather than a utilitarian substitute for the dingy sheet we've just converted into dusting rags. Unless sheets can become impulse items, manufacturers know their sales will simply be a reflection of the number of beds being sold and occupied. There isn't much growth in that.

It is difficult to sell sheets as a "want" item, making some sheets look or feel superior to "ordinary" sheets, at a greater price when department stores dangle irregulars on the covers of their advertisements for 50 percent or more off.

What is the purpose of the red tear string on Band-Aid brand adhesive bandage packages? Why did Nabisco eliminate the red tear string on the wrappers of its Saltine two-packs and four-packs?

If you have ever tried to open up a Band-Aid brand wrapper when treating a cut finger, you know how difficult it is to open up the package, retrieve the bandage, and apply it to the cut. In particular, if you follow directions and use the tear string, it will eventually slit the wrapper. But why bother? Why not just rip the paper?

The tear string was part of the original packaging of the product in the 1920s, and at the time, the tear string was state-of-the-art technology. The purpose of the string was to provide aseptic delivery of the bandage when sterility was essential, such as in surgical operations. A Band-Aid brand bandage is guaranteed to remain sterile as long as the package is intact, and the tear string allows the user to open up the wrapper without touching the bandage.

Consumers, however, who are most interested in ease and convenience, have not been wild about the tear string, and many simply rip up the paper wrapper without bothering with the string, even if this cruder method involves touching the bandage itself. Johnson & Johnson has finally responded to consumer preferences and is experimenting with different types of wrappers for some of its newer styles of adhesive bandages. Its Tricot Mesh and Handyman line come with paper wrappers without a tear string but are just as difficult to open as the regular packages.

Another new line, however, Flexible Fabric, offers revolutionary packaging. Rather than wrapped in semi-opaque paper, Flexible Fabric bandages come in sheer plastic, so it is possible, even in assortment packages, to know exactly what size bandage you are opening. Even better, the wrapper is a cinch to open. A piece of blue plastic hangs out, with the instruction, PULL. Using one hand to hold the backing, one simply pulls the plastic with the other hand and the package is opened. No need to touch the bandage, either. It's as simple to open as a Wrigley's chewing gum package.

A spokesperson for Johnson & Johnson Products Inc. confirmed that consumers prefer the new style. Although Johnson & Johnson bandages have become identified with the red tear string, it is likely that the Flexible Fabric style of wrapper will be extended to other lines, and perhaps the red tear string, which once was the best way to deliver a sterile bandage, will be retired.

Nabisco encountered much the same packaging problem as Johnson & Johnson. It had an excellent product, Saltine two-packs and four-packs, which dominated the market in restaurant crackers. Nabisco also wrapped its product with a red tear strip, and had since the late 1940s. But Saltines were even more difficult to open than Band-Aid wrappers. It was not uncommon to see frustrated diners rip open Saltine packages with their teeth.

Imponderables spoke with Bob Montgomery, of Nabisco Brands food service, who stated that although the red tear strip caused all kinds of problems, he resisted getting rid of it. The equipment used to wrap the Saltines gave Nabisco all kinds of problems, including snags on the high-speed production line. In theory, the red tear strip should have been easy for the consumer to grab by the fingertips—there was nothing theoretically wrong with the technology. But it just didn't work. Although there were not many consumer complaints (people don't tend to write to food companies about restaurant grievances), the red tear strip was slowing down Nabisco's assembly line, slowing down diners from consuming their crackers, and making their soup a lot cooler in the process.

The world of restaurant crackers has changed over the last decade. Whereas many restaurants used to provide four-packs, today about 80 percent stock two-packs and only 20 percent offer four-packs. In 1976, Nabisco eliminated the red tear strip. It had become so identified with the product, a symbol of luxury and quality, that Nabisco had good reason to be reluctant to give it up—many people, today, think that the two-packs and four-packs still have the red tear strip. That red tear strip signified that the restaurant was classy enough to provide a "name brand." Although Nabisco Saltine restaurant packs are now considerably easier to open (just rip the serrated edges—it's easy as pie), they've lost the embellishment that made their packages distinctive—a bittersweet trade-off.

Many clothing labels recommend against bleaching, yet many laundry detergents contain bleach. What gives?

Bleach does all sorts of nice things. It helps clean, whiten, and brighten clothes and works to remove soils and stains from fabric.

The two types of bleach that are used most often by consumers are chlorine bleach and oxygen bleach. Oxygen bleach, while providing all of the benefits listed above, has a gentler action than chlorine bleach; but chlorine bleach also disinfects and deodorizes and works faster than oxygen bleach.

Since it performs at least as well as oxygen bleach at all cleaning chores and also deodorizes and disinfects, most consumers prefer chlorine bleach as their laundry additive. But most detergents that include a bleach use oxygen bleach for one important reason—oxygen bleach is safe on any washable fabric. Chlorine bleach, stronger stuff, will wreak havoc on some sensitive but washable fabrics. If undiluted chlorine bleach comes in direct contact with some fabrics, tears and holes will result.

Many new clothing labels distinguish between the two bleaches when specifying "Do Not Bleach." When a label merely warns you off bleaching in general, be assured it is referring to chlorine bleach only.

Why do we eat ham at Easter?

The ritual of eating ham around Easter time predates Christianity. Ham was the entree at spring feasts for a very practical reason: Fresh meats were not available at the beginning of spring. So pagans buried prized fresh pork legs in the sand by the sea during the fall and winter. The pork was cured by the constant "marinating" of the salt water. The preserved meat was then cooked over wood fires. The pig was always a symbol of good luck and prosperity in the Indo-European culture, which made the choice of ham an even more felicitous choice for feasting.

In the early days of Christendom, Easter was celebrated on the Jewish Passover, and the two holidays are connected in several other ways. Just as Passover celebrates the Jews' release from bondage, so does Easter celebrate Jesus' release from the constraints of death. The paschal lamb, the ritual sacrifice made by Jews for each Passover, became emblematic of Jesus for Christians.

In ancient times, Jews ate the slain lamb for their Passover feast. They would never have contemplated eating ham, of course, since all pork products are proscribed by Moses in the Old Testament.

Early Christians, however, believed that Jesus' instructions superseded Mosaic law. While the Orthodox Christians had ordained fasts, they believed that no particular food should be prohibited from the diet.

Unfortunately, what started as Christian liberality in diet turned into some deliberate Jew-baiting. In particular, the English took to eating bacon at Easter in order to enrage Jews and repudiate everything Jewish. In the eleventh century, William (the Conqueror) I, who preferred ham to bacon, encouraged his subjects to make the switch. Pork, and particularly ham, has been the popular Easter meal in Western Europe ever since.

Not all cultures eat ham at Easter. In fact, England and to a lesser extent the United States have joined Eastern Europe and much of the Mediterranean in consuming lamb for their Easter feast. The choice of lamb, of course, derives from the Passover paschal lamb (although in Greece, for example, there needn't be any excuse to eat lamb). To some extent, one can successfully predict whether a given country will eat ham or lamb for Easter by determining which meat is consumed more every other day of the year.

The most likely explanation for the connection between eggs and Easter is that in medieval times, Christians were not allowed to eat eggs during Lent. So many eggs were accumulated during the forty-day fast that there were literally more eggs than they could eat. Why not decorate them?

An old legend gives a more moving explanation for the decoration of Easter eggs. Simon of Cyrene, an egg peddler by profession, was forced to abandon his wares and carry Jesus' cross to Calvary. According to Luke, when Simon returned to his basket of eggs after the Crucifixion, the eggs were unexplainably, and miraculously, beautifully decorated. According to this legend, we decorate eggs today to commemorate the memory of Jesus' sacrifice and the nobility of the humble Simon.

Why is film sold at 12, 20, 24, and 36 exposures?

According to the Eastman Kodak Company, the first Kodak cameras "were darkroom loaded, and depending on the size used rolls of film for 48, 60, or 100 pictures. When the first of the daylight loading roll films on flanged spools were offered, they came in 12-exposure rolls. Later, many of them were also made in 6- or 2-exposure rolls."

The Kodak Brownie, introduced in 1900, provided cameras for the masses, combining ease of use with a comfortable price. The camera sold for a buck; a roll of film for 15 cents.

While Eastman's flexible celluloid roll film eliminated the heavy and expensive glass plates that plagued casual users, the pictures were soft with inadequate detail; the serious photographer still could not find a combination of high-resolution pictures with portability. In the early twentieth century, the size of the camera was dictated by the size of the negative frame it held, until a man named Oskar Barnack revolutionized photography.

Barnack was an apprentice mechanic for an optical equipment factory and an amateur photographer on the side. He wanted to make larger pictures with smaller cameras, but he needed to find a way to reduce the size of the negative. Barnack was hired by the Leitz Company, which at that time was primarily a manufacturer of scientific instruments. Leitz was working on a motion picture camera, and Barnack hit on the idea of using motion picture film for his still camera.

Leitz shook the film industry by introducing the Leica camera at the Leipzig Spring Fair of 1925. Barnack created a camera that used 35-millimeter motion picture film and needed a mere 1 x 1½" negative. Not only was the picture quality much better than American cameras, but the motion picture negative was capable of almost limitless enlargements with relatively little loss of detail.

This Leica camera had a maximum capacity of 36 exposures. Originally, Leica customers had to load bulk film manually and insert it into the Leica camera. Kodak introduced preloaded cartridges in 1934, to fit the Retina, Contax, and Leica cameras, and Kodak adopted the Leica standard of 36 exposures as its length.

In 1936, Kodak introduced 18-exposure rolls but increased the size to 20 exposures for transparency films in 1946 and 24 exposures for color negative film in 1977. The increased size was a response to consumer preference for slightly larger roll sizes and not a result of any technological change.

The 12-exposure roll was first marketed by Kodak in Japan. Kodak introduced this size because of continuing interest in half-frame cameras, which make twice as many pictures on a roll of film. The 12-exposure roll was introduced in the U.S. because of consumer interest in a short roll. Although more expensive per picture than the larger rolls, the 12-exposure roll was the answer to the perpetual problem of the photographer harassing friends and relatives at parties by imploring, "Come on! Just eight more pictures! I want to finish off the roll."

When a building is on a corner of an intersection, how do they decide which street's name it will have?

If you were to build a department store in New York City and you had a choice between choosing 53rd Street or Fifth Avenue as your address, which would you choose? Fifth Avenue, of course, an address that conjures up images of elegance and affluence.

But who decides whether our mythical department store can receive its prestigious address? It's not the post office, which has nothing to do with assigning numbers or street designations to buildings. Usually, addresses are the domain of city or county offices. Geographers and engineers are hired to monitor the topographical planning.

In most cities, commercial buildings are handled differently from residences. In Manhattan, for example, most developers would prefer an Avenue address rather than a Street address. For some reason, real estate mavens consider Avenues ethereal and Streets as plebeian. The developer can receive his Avenue address *if the building has an entrance on that avenue,* even if the main entrance is on the other street and even if most of the property lies on the other street. Since many corner buildings have two entrances, the street address is essentially the owner's choice.

Developers are also fond of catchy numbers in their addresses. Their particular favorite is 1 (it gives the impression that their building owns the street), but round, even numbers, such as 100 and 10, are also popular, probably because they are easily memorized and signify that the building is big and important.

It is extremely common for small towns and industrial parks to accede to the wishes of companies in naming streets. At the very least, companies can nudge local governments into naming their byway "Park," "Plaza," "Court," or "Avenue" rather than the prosaic "Street." Many businesses like their address to be a walking advertisement. One of the most imaginative in this regard is Baskin-Robbins, whose Glendale, California, corporate address is 31 Baskin-Robbins Place. Most prestigious of all, perhaps, is when a corporate address is too rarefied and important to require something as piddling as a number: Campbell Soup can be reached at Campbell Place, and Kraft, Inc. at Kraft Court.

If a building is large enough to occupy several city lots, it can be given any of the numbers ordinarily assigned to those spaces. Assume there are five existing buildings on a block,

110, 120, 130, 140, and 150. If a new building that is the same size as the other five combined is constructed on the corner, it could become 160, 170, 180, 190, or 200. City engineers might prefer the building to choose 180 since it best describes its location, but the building's owners would almost certainly prefer the more memorable 200.

Residential owners in most localities do not have a choice over their street designation. Most local governments issue addresses based upon where the front entrance is located, even if the bulk of the house lies on the other street. If an owner builds a second entrance facing the other street, an application can be made to change the address of the house. It is possible then, if the entranceways are so located, that all four corner residences at one intersection might carry the same street name.

What does ½ at the end of a street address mean?

As the above Imponderable indicates, address numbers are assigned with a great deal of flexibility. On densely populated blocks, each building is given a number at least two away from its neighbor. But what if 116 Main Street is subdivided and 114 and 118 Main Street are already taken? If a new owner wants a separate address, there is nothing for city planners to do but give the new edifice the old address, 116, plus one-half, or 116½ (in some localities, the old address becomes 116A, the new one, 116B).

Contrary to popular belief, a ½ designation does not necessarily mean that a building is small or is owned by the same person as the building with the same number minus the ½. Although city planners dislike using fractions in addresses, it is the only solution for numbering new buildings built between two existing edifices that are only two numbers apart.

Why do wintergreen Life Savers sparkle in the dark when you bite into them?

Be the life of your next party. Buy a few rolls of wintergreen Life Savers Roll Candy, cut all the lights, gather your friends in front of you, and bite down hard and fast. You'll sparkle in the dark. Your mouth will glow bluish-green.

The explanation for this delightful phenomenon comes directly from the research and development department of Life Savers Inc., which is now a division of Nabisco Brands:

> Our manager of Candy Technology tells us that two ingredients are necessary for this reaction. The sparkling comes about because of a combination of mint flavoring and crystalline sugar. When you crack the crystal, the energy then stimulates a component in the flavoring to emit a light. The component in wintergreen is methyl salicylate.

There are two possible hang-ups in producing the sparkling effect. First, the background atmosphere might not be dark enough (closets and bathrooms are highly recommended). Second, moisture seems to absorb the energy needed to produce sparkling. Do not expect good results in a sauna.

Why do labels usually tell you that top-loading washing machines require more detergent than front-loading washing machines?

Have you ever noticed, as Andy Rooney likes to say, that you have to dump twice as much detergent into a top-loader as a front-loader? And that somehow, *you* always seem to be using a top-loader? Is there such a thing as a front-loading washer? Do top-loaders really require more detergent or are detergent executives taking retreats in the Bahamas because we are dumping in needless cupfuls?

We could answer these questions right away, but that would be too easy. For there are many factors besides whether your washing machine is a front- or top-loader that determine how much detergent you should use, and none of them are specified on detergent packages. You need to know the hows and whys of detergent labeling to enhance your detergent I.Q. You need to know, as Paul Harvey likes to say, the rest of the story.

You'll never know a blasted thing about laundering until

you learn about what we cognoscenti call the Big Three. The Big Three are the requirements for cleaning performance, and without further ado, these are the types of action needed to clean anything from lingerie to mattress pads: mechanical, thermal, and chemical. The neat thing about the Big Three is that they are totally interrelated. If any one is reduced in energy, the others must be increased to achieve the same level of cleaning. If one is increased, the other two may be proportionately lessened without hurting performance.

Mechanical Action

Letting soiled clothing soak in plain lukewarm water will result in wet, lukewarm soiled clothes. Some form of friction must be exerted to pull off dirt. Before electricity, women pounded rocks against clothes to loosen soil. Front-loading machines spin the wash around, much like dryers do. Top-loaders, with agitating action, provide more mechanical stimulus more efficiently. If the other two components of the Big Three are lessened, it is possible to achieve slightly better performance if the wash is agitated for longer periods of time, but this is the most difficult of the Big Three on which to gain performance.

Thermal Action

The temperature of the wash cycle is perhaps the most crucial element in determining cleaning effectiveness. The evidence is unanimous. Hot water cleans better than cold water. You might not *need* to use hot water to clean lightly soiled items, and certain items are too delicate for hot-, or even warm-water washing, but there is no intrinsic cleaning advantage in cold water. Cold water washing became popular not because of any claims for its performance superiority, but because of its energy savings.

Heat is essential to combat stubborn oil or greasy soil. Most experts recommend setting water heaters to 140 degrees. The actual temperature of the water in the washing machine will be at least ten degrees cooler due to the effect of the plumbing and contact with the room-temperature garments in the wash-basin.

If you use the cold water setting on your washing machine, you are getting whatever temperature water comes out of the cold side of your tap. In Florida during the summer, the "cold" water might be 90 degrees; in Minnesota during the winter, 35 degrees. When the temperature of the water is less than 60 degrees during the wash cycle, there is hardly any cleaning action. What will happen if you use a cold-water detergent with water under 60 degrees? Not much. One of the main problems is that most powdered detergents don't dissolve well in temperatures below 70 degrees—even those touted as cold-water detergents.

There are ways to compensate for the weaker performance of cold-water cleaning. Dissolve powdered detergents before using them. Not only will this unleash the full cleaning action of the detergent, it will also eliminate disgusting gunk from clinging to the clothes (remember Salvo, the convenient powdered cake that cleaned so well but refused to dissolve?). Another tip is simply to use more detergent—up to twice as much as the label of a regular detergent specifies—when using cold water. But the best solution of all, unless your clothes are too delicate, is to use warm or hot water to wash your clothes and use cold water for the rinse cycle. A cold-water rinse is just as effective as a warm rinse, and the hot wash and cold rinse will cost about as much money as a warm wash and warm rinse.

Chemical Action

The more cynical among us might assume that there is no difference between bar soap, bubble soap, dishwashing liquid, shampoo, automatic dishwashing liquid, and laundry detergents. The cynics are wrong. Laundry detergent is designed for its specific task. A true chemical action is necessary not only to loosen and remove soil from clothing but to suspend the soil, lint, and grease in the wash water until they are drained.

Soap's effectiveness is greatly diminished in hard water. Many common mineral salts, particularly calcium and magnesium, but also iron and manganese, cause hardness. Hard water tends to turn soap into insoluble curd (a.k.a. the "Salvo Crud"), which will not rinse away.

During World War II, when oils and fats were in short supply, artificial detergents were hastily developed, and by 1953 artificial detergents, led by Procter & Gamble's Tide, supplanted soap as the United States' favored laundry-cleaning agent. Detergents were not only more effective in cleaning than simple soaps, they also relieved many of the hard water washing problems, since they are chemically built to resist hard water minerals.

Detergents are composed of two types of ingredients. Surfactants (surface active agents) are a compound with two ends, one that repels water and the other that attracts water. Surfactants facilitate dispersion of water onto laundry and concentrate at the interfaces between soil and water and between water and fabric. This action solubilizes the soil and removes it from the fabric. The other type of ingredient in all detergents, the builders, help to soften the water by deactivating the hardness minerals. Builders break oil and grease into tiny globules, suspending dirt in the water until it is drained away. Builders also provide the desired level of alkalinity in wash water, which aids in cleaning. Without surfactants and builders, you would need considerably more detergent to achieve the same level of cleaning, which is why natural soap detergents, although cheaper, require more detergent per load. Even with builders, however, hard water requires more detergent than average or soft water.

The only rub with modern detergents is that those remarkable builders, which soften the wash water, are usually phosphate-based. Six states have outlawed phosphates entirely in detergents, for environmental reasons. Phosphate is cheap, and this was the crucial ingredient in making chemical formulations cost effective and cleaning effective more than forty years ago. Without phosphates, we have a better environment but an old cleaning problem. Detergents must either be expensive (the chemical alternatives to phosphate are not cheap) or relatively ineffective at hard-water cleaning. More detergent must be used, and because nonphosphate detergents run into solubility problems, about half of all consumers in nonphosphate states use heavy-duty liquids rather than detergents, about double the national average (liquids generally outperform powders in all hard- or cold-water cleaning).

Although top-loaders are gaining ground, front-loading washing machines are still prevalent in Europe. Most of them have revolving drums designed to conserve energy and water. Because the water spins with the clothes, European front-loaders use less water in relation to fabric than top-loading machines. Without the agitator, however (the mechanical part of the Big Three), Europeans compensate by increasing the length of all of the wash and rinse cycles and by using hotter temperatures than American consumers. Aha, thoughtful readers might point out, couldn't the lesser mechanical action of the front-loader be compensated for by increasing the chemical action (i.e., using more detergent)? If the mechanical action of the front-loader is less than the front-loader, shouldn't the front-loader use *more* detergent?

Theoretically, yes. But the tumbling action of European washers tends to generate excessive foam, so their detergents contain chemical ingredients designed to suppress suds. With a low-sudsing detergent, it is safe to use the same amount of detergent for both types of machines. But as Molly A. Chillinsky of the Coin Laundry Association put it, "A high sudsing detergent in a front-loader would cause the soap foam to literally fall out onto the floor, a situation not particularly desirable to most launderers." Nor to most laundromat owners (or managers), the biggest customer for front-loaders. With front-loaders rarely being sold anymore, most detergents don't even specify whether or not they are high- or low-sudsing products. Still, the danger of "fallout" is one of the reasons detergent labels specify less detergent for front loaders.

Remember a few eons ago when we mentioned that the

answer to this Imponderable was "too easy." It's time to fess up. The main reason that top-loaders require more detergent than front-loaders is because they have bigger capacities. Top-loaders can hold more water. Top-loaders can hold more clothes. Front-loaders usually hold between 8 and 10 gallons. Top-loaders have a capacity of 10 to 25 gallons.

When a detergent label recommends using a certain amount of its product, it is assuming average conditions. Although producers never specify what "average" means, here are their implicit assumptions:

1. An average-size load (approximately five to seven pounds of clothes).

2. Average dirtiness (i.e., no heavily soiled garments). If you are using a liquid detergent and have an average-size load with one or two heavily messed-up garments, pretreat the stains with your detergent, but add the regular full dose to the wash as well. Do not include detergent used for pretreating as part of your detergent allotment for the wash. If you don't know what less-than-average dirtiness is, neither do we, but the best rule of thumb is that it is generally better to use too much detergent rather than too little—at least that's what the detergent companies told us for some reason or another.

3. Average hardness of water (consumers presumably use the Federal Water Hardness Hotline to determine the relative hardness of their water).

4. Sixteen gallons of water for a top-loader, eight gallons for a front-loader (which is why most labels tell you to use twice as much detergent in a top-loader as in a front-loader). Skeptics feel that the only reason the companies retain the instructions for front-loaders is so that they can promote in advertising and in big print on their packaging that consumers need only use one-quarter cup of their detergent. The companies needn't add that it would be highly unusual if the customer roaming the grocery store aisle happened to be lucky enough to own an obsolete washing machine.

To be fair, some detergent manufacturers have stopped printing their two-tiered instructions, and *Imponderables* tried to find out why. Lever Brothers guessed that it may be because so few front-loaders are manufactured and sold anymore. Molly Chillinsky had a different hypothesis. In some states, a few localities have banned phosphates, so manufacturers are virtually forced to distribute two different versions of their detergents. "As a result, directions for proper amounts of detergent to be used would be determined by that factor, rather than by which type washer the consumer is using."

What does the "L.S." next to the signature line on contracts mean?

Next to the signature line on most formal contracts, you will find the cryptic abbreviation L.S., usually in parentheses. Many people assume that L.S. stands for "legal signature." It doesn't.

L.S. is an abbreviation of the Latin, *locus sigilli*, meaning "the place of the seal." Centuries ago, seals were usually essential in order to make a document or contract official. A seal, according to *Black's Law Dictionary*, could be "a design, initial, or other device placed on a letter, document, etc., as a signature of proof of authenticity." Letters were once closed with a wafer of molten wax, leaving a distinctive, indelible mark of the sender.

Nowadays, a simple signature usually will suffice. Although the federal government and all fifty states have their own seals and many corporations and government departments have them as well, a naked signature on a letter or contract is usually no less binding than one with an ornamental seal.

Many contracts still insert the L.S. next to the signature line even though it is an anachronism likely to disappear eventually. While it once marked the *place* of the seal, L.S. now indicates that the signature is *instead of* the seal.

What's that funny beep on the radio just before the network news?

You are hearing a two-frequency tone known as a bee-doop. The bee-doop is an electronic means of communication from a radio network to its individual affiliates. Each affiliated station is equipped with encoding equipment that translates the meaning of the particular bee-doop sent by the network.

One bee-doop might tell the affiliate that the network news is airing immediately. One might notify that a live sports feed is about to be sent. Another might alert the affiliated stations about a national news bulletin that might interrupt local programming. Most networks use a bee-doop to broadcast closed-circuit programming and promotional announcements to their stations. The local stations can set their encoding box to receive the bee-doop as it is sent or as a flashing light with no audio.

The most important use for the bee-doop is in signaling automated radio stations. Many radio stations, particularly in small markets, are fully automated, with only a lonely engineer as the sole employee at the studio. These bee-doops signal the encoding boxes to switch from local to network programming and to automatically trip recorders to tape network feeds for rebroadcast in the local market.

Bee-doops could serve an important function in case of national emergency as well, since they can act as an instantaneous communication device among hundreds of stations in all fifty states. George Thomas, the manager of broadcast operations for the Mutual Broadcasting System, told *Imponderables* that MBS tests the Emergency Broadcast System through its internal bee-doop system. The President, in case of an emergency, could reach many stations (since a large number of stations have a network affiliation) simultaneously this way even if some regional outlets for the Emergency Broadcast System were knocked out.

Although bee-doops have become a subliminal listening experience for radio audiences, most affiliates don't like them, feeling that they make their stations' broadcasts sound unclean and their programming transitions too awkward. Most of the radio networks have therefore reduced substantially the volume of their bee-doops.

Why are cities warmer than their outlying areas?

In almost every metropolitan area in the United States and Canada, the city is warmer than its immediately surrounding areas. Compared to suburban and rural areas, cities have gotten warmer throughout the twentieth century.

Do the cities themselves generate enough heat to raise the temperature measurably? Is there something about cities that allows them to retain heat? The answer to both questions: Yes.

The heat generated by buildings, factories, vehicles, lighting, and other byproducts of modern technology is enough to raise the temperature a degree or two in densely populated cities. The hot air exhaled by air conditioners during summer months affects the temperature outside as surely, if less dramatically, as it affects the temperature inside an air-conditioned room.

But even if cities did not generate their own heat, they would still be warmer than rural or suburban areas. When the sun shines on the flat, featureless Kansas countryside, the light is reflected back to the sky. When the sun shines in midtown Manhattan, the light bounces from skyscraper to skyscraper like a manic Ping-Pong ball—more of the sun's warmth lingers close to ground level than on the Kansan farm and more warmth is absorbed in the city. In fact, buildings and cement pavements can retain more heat and more sunlight than grass, trees, or the farmer's topsoil.

Precipitation has a cooling effect in the country. Rain is stored in the ground and recycles itself through evaporation and plant respiration, thus absorbing heat. In the city, precipitation is funneled into sewers, effectively eliminating much of its cooling effect. The relative lack of this evaporation in the city explains why cities tend to be less humid than rural areas.

It is commonly assumed that air pollution is what makes cities warmer. Since dust particles can absorb radiation, the theory goes, the more polluted the city, the higher the temperature is artificially raised. There is only one problem with this hypothesis: Dust particles can also *reflect* radiation, bouncing rays that would otherwise be trapped near ground level back up to the sky. The jury is still out on the net effects of pollution on temperature.

One fact remains indisputable, though. On extremely windy days, the temperature differences between city and country tend to disappear; on calm days, there is more of a discrepancy than normal. The wind mitigates human intrusion upon the "natural" climate.

While modern life hasn't seemed to affect wind patterns, we have already created a lifestyle that might permanently change our temperature patterns, at least in metropolitan areas. Meteorologists have little idea, at this point, if these barely perceptible changes (cities have become a few degrees warmer in the last fifty years or so) will create profound changes in our ecosystem. They might. And we could usher in the next Ice Age with our cities as hothouses.

What is the purpose of the little slit in the folds of sugar cube wrappers?

Imponderables asked the wrong question. It turns out that *there is no purpose* for the little slit in the folds of sugar cube wrappers. But there is *a reason* for them.

The tiny wrappers are die-cut and put in a magazine, where they wait until they contact the sugar cubes. In the past, because the paper was so thin, machines used to grab six or eight wrappers at a time, wasting money and producing some pretty silly packages. The solution to the problem was a pin to hold down the stack of wrappers and then a needle to pick up each label individually. When the needle lifts each wrapper, it rips the label slightly while pulling it out of the stack. The slit serves no function, but is the byproduct of a simple but efficient system of machine packaging.

Unfortunately, it is getting harder and harder to find sugar cube wrappers, let alone their little slits. The sugar cubes you find in stores are no longer wrapped, and most restaurants have abandoned the cube for crystalline sugar. Crystalline sugar is a more flexible product, enabling the consumer to pour exactly the preferred amount of sugar into coffee or cereal bowls. But there is only one reason why sugar cubes have disappeared from restaurants—cost. Domino Sugar manufactures crystalline sugar packets in the billions every year, but wrapped cubes only in the hundreds of thousands. It costs restaurants several times as much per serving to buy wrapped sugar cubes, so the few who still serve them tend to be exclusive restaurants, with their own logotypes affixed to the wrapper. Some ethnic restaurants also feature sugar cubes. Many first-generation Eastern European and Mediterranean immigrants retained the custom of drinking coffee with a sugar cube in their mouths; others dunked whole cubes into coffee and sucked on them. But these immigrants' children are not following the traditions. The sugar cube, slit and all, is in danger of extinction.

Why does the word *Filipino* start with the letter *F*?

The Philippine alphabet does not contain the letter *F*. Natives refer to themselves as Pilipinos. This spelling, not phonetic in English, led us to use *Filipino*, the Spanish word for the people over whom they once ruled.

Although the confusing English phonetics of Pilipino explains why we use an *F*, it is still surprising that we didn't coin our own word, *Philipino*, which would retain our spelling of the country, but the *Oxford English Dictionary* notes that the word *Filipino* was used in American newspapers as early as 1898.

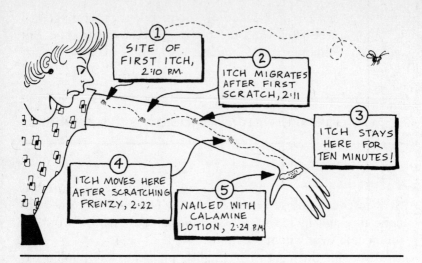

Why do we itch?

The short answer is: We don't know.

Here's the long answer. Itching is an enigmatic phenomenon. If a patient complains to a doctor that she has horrible itching and the doctor finds hives on the surface of the skin, the doctor can treat the growth and alleviate the itching symptoms. But much itching has no obvious cause and is not associated with any accompanying illness. Scientists can induce itching by heating the skin too close to the pain threshold or giving subjects certain chemicals, especially histamines (thus explaining why doctors prescribe antihistamines as a treatment for itching), but the ability to induce itching doesn't mean that doctors know its etiology.

This much is known. There are sensory receptors just below the surface of the skin that send messages to the brain. The itch sensation seems to flow along the same pathways of the nervous system as pain sensations. According to Dr. George F. Odland, professor of dermatology at the University of Washington Medical School, the vast majority of sensory receptors are "free" nerve terminals. These "free" terminals do not seem to be designed for any specialized or particular function, but they carry both pain and itch sensations to the brain. These pain receptors are the most common in our nervous systems. When they operate at a low level of activity, they seem to signal itchiness rather than pain.

Many scientists have speculated about the function of itching. Some believe that itching exists in order to warn us of impending pain if action is not taken. Others speculate about the usefulness of itching in letting primitive man know it was time to pluck the vermin and maggots out of his skin and hair. Itchiness can also be an early symptom of more serious illnesses, including diabetes and Hodgkin's disease.

Itching sensations are distinct from ticklishness, which at least some people find pleasurable. Itching is rarely pleasurable; in fact, most people tolerate itching less well than pain. Patients with severe itching are invariably more than willing to break the skin, including pain and bleeding, in order to remove the itch.

Why do gas stations use machines to print out the amounts of credit card purchases when other merchants write out the numbers by hand?

The major oil companies, like Mobil, Exxon, and Texaco, receive literally millions of credit card slips a day. It's no wonder that these corporations have depended upon automation to speed up the processing of these slips.

Even when their total daily gross is less than other types of businesses, gas stations tend to issue more credit slips per day than any other kind of establishment—more than most restaurants, hotels, or retail stores. Gas companies have historically encouraged credit card sales, believing that a credit card holder will tend to remain loyal to the issuing company. A spokesperson for Mobil estimated that 30–35 percent of its total volume is from credit card purchases. The service station owner does not necessarily lose cash flow when you buy with a credit card. Most oil companies allow service stations to buy their gas and other products from their home company with cash, check, or credit card slips, so the station owner need not wait for you to pay the bill before the station is reimbursed.

Optical scanning devices became prevalent in the oil industry in the late 1950s and early 1960s. In the good old days of full service, the gas jockey would take your credit card, put it into an imprinter, and move three or four levers to print out each digit in order to encode the price of your purchase. In most service stations today, the technology is the same. With the amount of the purchase printed in purplish numbers, optical scanning devices can read the amount without the need for a keypunch operator to examine each credit slip individually. Although optical scanning devices are extremely expensive, current ones can read slips at a rate of 100 per second, which obviously saves time and money. With the new point-of-sale electronic-magnetic recorders, imprinters are going to become merely backstops in case of problems with the scanning device, since the new technology greatly reduces the amount of time service station employees must spend filling out a credit card slip.

Some optical scanners are theoretically capable of "reading" handwritten numbers, but at present they are inconsistent in performance. Until these types of scanners are perfected (and brought down in price), the oil industry is content to stick with technology that is reasonably priced and reliable.

So if the gas stations have such a great deal going, why don't Visa, MasterCard, American Express, and the other big boys come along for the ride? An oil company representative mentioned that one of the initial reasons for using imprinters was that gas stations were less than pristine places, and there were problems with humans, let alone optical scanners, reading numbers with oil smudges on them. Perhaps, he theorized, credit card companies hadn't shared their problems in reading the credit slips and thus had no pressing need to utilize scanning technology. Another oil executive proposed the idea that perhaps the volume of credit card slips processed by even the biggest credit card companies was not sufficient to employ optical scanners.

Both theories were wrong. The major credit card companies *do* employ optical scanners, which read all of the information on the slips except for the amount. There are two main reasons the major credit card companies have merchants manually fill out credit card slips. The first, and most important reason, is that many credit card purchases are made in places where a gratuity might be added (in the case of American Express, about 30 percent of the places where a card is used). The credit card companies want to do everything they can to encourage their customers to put the tip on the credit card, since they will collect a percentage of the gratuity as well as the purchase price. If gratuities were added after the original price of, say, a gourmet dinner, it would require two credit slips or tearing up the original slip and imprinting another—in either case necessitating extra work for the merchant and extra time and inconvenience for the patron. Visa, MasterCard, and especially American Express extract their fortunes by eking out huge profits from tiny profit margins (after all, traveler's checks are sold with a 1 percent markup). Losing their percentage of gratuities would offset any gain from using a fully automated point-of-sale device.

The second reason the major credit card companies might be reticent about using optical scanners is that while the average oil company bill might total less than $20, many credit card purchases on all-purpose credit cards can come to thousands of dollars. If a scanner happens to drop a zero or miss a decimal point, it takes a lot of savings from the wonderful world of automation to compensate for one colossal automated boo-boo.

What is the difference between "gourmet" popcorn and "regular" popcorn?

Different species of popcorn have the same flavor, but food technologists can fiddle with other characteristics that make popcorn more appealing to the consumer. There are only two basic types of popcorn: yellow and white. Yellow popcorn always pops bigger than white and is therefore favored by concessionaires, since it fills up more buckets with no more outlay for kernels. Many popcorn aficionados prefer the white variety, which has a slightly sweeter taste.

"Gourmet" popcorn is a hybrid. While retaining some of the extra sweetness of white popcorn, "gourmet" popcorn is bred to pop bigger than ordinary yellow kernels. The degree to which a burst popcorn kernel exceeds its original size is known in the popcorn trade as its expansion ratio. "Regular" popcorn averages an expansion ratio of 35–38 times, but the best hybrids average over 40. Makers of "gourmet" popcorn, such as Orville Redenbacher, also claim that their product is less tough than regular popcorn, providing a "melt in your mouth" consistency that consumers covet.

Why aren't cashews ever sold in their shells?

Cashews aren't sold in their shells because they don't have a shell. Don't all nuts have shells? Yes. Then what gives?

Imponderables would never be pedantic, but we must insist that a cashew is a *seed*, not a nut. The cashew is the seed of a pear-shaped fruit, the cashew apple, which is itself edible. The cashew seed hangs at the lower end of the fruit, vulnerable and exposed. Cashews grow not on trees, but on tropical shrubs, similar to sumac plants.

A hard or leathery shell is what differentiates a nut from a seed. Kernels with thin, soft shells, such as pumpkins and sunflowers, are properly called seeds.

Why do the minute hands on school clocks click backward before advancing?

How we remember the torture inflicted on the boy in the above cartoon. During the last period, we waited impatiently for that last tick of the clock that would bring the school bell that would signal our freedom from school. And just when we heard that jolt of electricity hit the clock, the minute hand would go backward instead of forward. For a split second, we feared that our principal had figured out how to reverse time.

Our paranoia was heightened by our knowledge that the master clock—which sent the impulses to the clocks in our classrooms—was housed in the principal's office. The master clock sends commands to the secondary clocks located throughout the school to advance every minute, sending an electrical charge that lasts approximately two seconds.

The back-tick of the minute hand is the result, not the purpose, of the design of impulse clocks. John Evans, of the Lathem Time Recorder Co. in Atlanta, Georgia, was kind enough to draw a diagram of the back of a secondary impulse clock, the

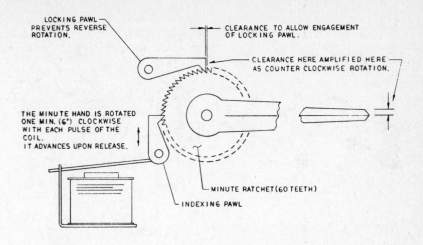

LOCKING PAWL PREVENTS REVERSE ROTATION.

CLEARANCE TO ALLOW ENGAGEMENT OF LOCKING PAWL.

CLEARANCE HERE AMPLIFIED HERE AS COUNTER CLOCKWISE ROTATION.

THE MINUTE HAND IS ROTATED ONE MIN. (6°) CLOCKWISE WITH EACH PULSE OF THE COIL. IT ADVANCES UPON RELEASE.

MINUTE RATCHET (60 TEETH)

INDEXING PAWL

type that would be in the average classroom. The ratchet wheel that you see in the diagram moves one tooth at a time, which assures that the minute hand on the face of the clock moves precisely six degrees every minute. To ensure that the minute hand moves properly, a locking pawl also engages the teeth of the ratchet wheel to prevent a reverse rotation. For the locking pawl to do its job, it must momentarily allow clearance for the wheel to move while the coil is being energized—otherwise, the wheel couldn't advance one tooth at a time. While the clearance is in effect, the locking pawl drags the ratchet back a tiny bit (perhaps as little as one-tenth of a degree). This effect is magnified, however, many times over by the movement of the much larger hands of the clock.

Most synchronized clock systems are self-correcting. Usually, they can make minor corrections of a minute or so on an hourly basis, often at one minute before the hour; when a classroom clock is running too fast and gets corrected at 2:59 P.M., school kids are tortured even more. Major corrections are executed automatically every twelve hours (usually, in schools, at 6:00 A.M. and P.M.). Some school systems have computerized their clock systems, with the principal's master clock automatically tripping off the school bells.

If only some people are susceptible to hypnosis, how can stage hypnotists confidently ply their trade?

While on one of our extremely rare breaks from work on this tome, we happened to flip on the television and watch a nationally syndicated talk show. The special guest was a hypnotist, who proceeded to put three members of the studio audience "under." The three subjects, all women who wanted to quit smoking, were put in a trance. The hypnotist told one of them that when he signaled her to wake up, she would have an urgent craving for a cigarette, but when she smoked it, it would taste like common sludge.

Like a trained seal, the subject, when awakened, furiously lit a cigarette and, after inhaling, did a take with as much finesse as a consummate comedienne. What had been the object of her desire only a second ago now repulsed her. The very brand of cigarette she chain-smoked now tasted like a combination of chalk and paste.

You've probably seen other stage hypnotists at work. Perhaps the most famous is Pat Collins, the "Hip Hypnotist," who has starred in many Showtime specials and in her own nightclub on the Sunset Strip. She can effortlessly transform seemingly normal spectators into would-be opera singers and burlesque stars.

Textbooks on hypnotism assure us that not everyone can be hypnotized and that even among those who can, there is a wide range of susceptibility and suggestibility. Yet stage hypnotists have been popular entertainers for over a century, relying on unknown volunteers from the audience as their subjects, and one rarely sees a hypnotist fail at his or her appointed task. When we see a stage magician, we assume trickery. We buy our ticket specifically to see trickery. Is stage hypnosis fakery in the same sense, or do stage hypnotists know something clinical hypnotists do not?

To find out more about this Imponderable, we used many sources. Not surprisingly, no stage hypnotists would talk to us on the record—or much, for that matter, off the record. But we unearthed quite a bit of material written by stage hypnotists for their fellow practitioners, as well as enough medical and scholarly material to piece together much of the story.

Most skeptics believe that stage hypnotists use confederates who pretend to be hypnotized. Actually, in the days of burlesque this was sometimes the case. Most hypnotists, even then, preferred to find confederates who could be put into a trance, for their most popular tricks involved physical stunts, such as placing the subject's head and feet on separate chairs and having the subject's body lie as rigid as a corpse. Why didn't the subject's body collapse without support? Because these confederates were predisposed toward falling into hypnotic catalepsy, a condition in which consciousness is temporarily lost and the muscles become rigid. Epileptics, during severe seizures, are cataleptic, and some schizophrenics during psychotic episodes are similarly impervious to pain and suffering. With a cataleptic confederate in hand, at least one segment of the entertainment was consistently foolproof.

Other bogus hypnotists used props to achieve their effects. Some "stopped the blood from flowing" by placing a golf ball under the armpit of the subject. The confederate simply pressed his or her arm against the ball and circulation suddenly stopped.

These crude tricks might have worked for barnstormers who performed one-nighters in small towns, but they are dangerous and needlessly expensive today. Unless the hypnotist can find a different stooge at every performance, the confederate runs the risk of being discovered, discrediting the reputation of the hypnotist. Even the best confederate is a financial liability—an unnecessary mouth to feed. Props are also dangerous, especially in television performances, where close-ups are a fact of life. Stage hypnotists don't need confederates or props to successfully ply their trade, but they do need to follow some of the procedures described below to ensure a successful act.

Find the Right Subject

Most stage hypnotists call for volunteers from the audience. If the subjects onstage were a random sample of the general population, the hypnotist would be in trouble. But they aren't, for several important reasons:

1. *Volunteers are self-selected candidates to be good subjects.* People who go to a hypnotist for treatment are more likely to be hypnotizable than the average person, if only because their mere presence confirms their belief in the efficacy of hypnosis. Attendees at hypnotic demonstrations are even more susceptible, and volunteers, except for the professional skeptics, who deliberately try to undermine the stage hypnotist (more on them later), are likely to be extremely hypnotizable.

Psychologist Martin T. Orne points out in an article in the *Journal of Abnormal Psychology* that ". . . much hypnotic behavior results from the subject's conception of the role of the hypnotic subject as determined by past experience and learning, and by explicit and implicit cues provided by the hypnotist and the situation." To extend Orne's thesis, if a subject, before going onstage to volunteer, believes in the power of the hypnotist, presumably because of past exposure to or experiences with hypnosis, the subject is going to be more willing to suspend customary behavior and, perhaps, more importantly, is going to feel no responsibility for his or her behavior onstage when "under" hypnosis.

2. *An effective stage hypnotist chooses subjects by carefully determined criteria.* Almost every hypnotist does a "demonstration" before commencing his or her act. Usually, the ostensible purpose of the demonstration is educational—to inform the audience a little bit about how hypnosis works or to allay audience fears about what the hypnotist will maintain is not a mystical process. The actual purpose of this demonstration is to search for susceptible prospects for the act.

Most hypnotists use a couple of old chestnuts for this demonstration. They may put the audience as a whole "under" and inform them they have just eaten a tremendously sour lemon. By carefully scrutinizing the reactions of an audience *with its eyes closed*, the hypnotist can easily spot promising prospects. Even more commonly, the hypnotist will ask audience members to squeeze their eyelids together. Then the audience is informed that their eyes have been shut and cannot be opened. The hypnotist asks the audience to try to open their eyes. Those who can't open their eyes have passed the test.

The hypnotist is likely to socialize a bit with the crowd, encouraging verbal interaction, which serves to loosen up the audience for the act, but more importantly allows the hypnotist to observe other behavioral traits that determine successful volunteers. In general, the hypnotist is looking for highly sociable people, uninhibited types who will not fear looking foolish in front of other people. People who are terrified at the prospect of losing control of their emotions are the worst candidates for hypnosis. The hypnotist wants to find friendly but trusting types, who will be willing to subordinate their egos while onstage. These psychological factors are as important in scouting prospects as the lemon or eye tests, and audiences rarely have an inkling that the hypnotist's repartee is masking crucial preparatory work. Although you won't see this part of the act on a television program, stage hypnotists use this type of warm-up on talk shows before proceeding with their act—it is a crucial diagnostic tool for the hypnotist in screening subjects.

The hypnotist is faced with one more major problem in selecting subjects. Even if the hypnotist discovers the best prospects in the audience, it doesn't mean they will volunteer. Folks tend to be skeptical when the hypnotist selects subjects; they are much less suspicious when volunteers are sought. Furthermore, there is another advantage to selecting "real people" from the audience—they will be friends of others in the crowd, heightening audience interest and enthusiasm.

So how does the hypnotist get to work on the most susceptible audience members while supposedly asking for any volunteers? Obviously, if enough people volunteer who passed the pre-tests, the hypnotist should select them. If some of the most promising prospects are recalcitrant, the solution is to allow volunteers to come onstage, but to perform difficult tricks only on those who have shown a disposition toward susceptibility. Several manuals for stage hypnotists suggest that the hypnotist diplomatically but firmly approach those who are not susceptible and simply ask them to leave the stage. Many hypnotic stunts do not require the entire group of subjects. In the talk show we watched, for example, only one of the three subjects was used for the most demanding stunt. When group stunts are needed, the hypnotist has many tricks in order to induce subjects to conform to the behavior of the rest of the group (see below).

Pressure the Subjects to Behave in the Desired Manner

The hypnotist is at a tremendous advantage over the subject during the show. A volunteer mission is a submissive one. It is amazing that so many people are willing to go onstage and risk losing control in front of hundreds (or in the case of television) millions of people. Yet, there is no problem hiring recruits for shows like *Let's Make a Deal* or *Truth or Consequences*, the latter of which didn't even provide much financial compensation for fans willing to humiliate themselves. It is impressive what good sports the "bargainers" are on *Let's Make a Deal*, dressing and acting outlandishly in order to attract Monty Hall's attention and then smiling as they trade boats they have won for a basket of hard-boiled eggs. The same contestant might cry buckets if she lost a wallet with fifty dollars in it. It is unclear whether game show contestants are in a trance, but hypnotists have many tricks to make their subjects as compliant as Monty Hall's contingent. Here are some of the pressures, usually exerted by the hypnotist, that encourage volunteers to become docile subjects.

1. *The volunteers should be made to feel that the success of the act weighs on their shoulders.* Unlike magicians, who use volunteers as comic foils, the hypnotist makes sure the subjects know how important they are. This places tremendous pressure on subjects, since they tend to be nervous to begin with and have now been told that the hypnotist is dependent upon them.

2. *The alien world of the stage, especially with lights, cameras, microphones, etc., places the subjects in a vulnerable position.* And more likely to feel dependent upon the hypnotist.

3. *The audience, unwittingly, acts as a major source of pressure on the subject, and a good hypnotist will intensify this phenomenon.* Much of the time while onstage, subjects have their eyes closed. Their only sources of information are the voices of the hypnotist and the studio audience. With their laughs and applause, the audience rewards "good" behavior and punishes, if only with silence, "bad" behavior. Usually, "good" behavior is defined as outlandish behavior, and subjects, who are by no means incapable of sensing audience reaction, want to please the audience as well as the hypnotist (and many hypnotists get laughs from the audience by making sarcastic comments about reticent subjects, making it even harder for subjects not to conform to the wishes of the hypnotist). Of course, exhibitionists love to perform in front of an audience and feel the same performance pressure that professional actors would. Hypnotists love "hams,"

since any bizarre behavior they exhibit will be credited to the hypnotist rather than the subject. One ham can nullify the harmful effects of a stage full of duds.

4. *The hypnotist's opening remarks subtly indoctrinate the audience into believing the idea that hypnotizable people are superior types.* During the hypnotist's introductory lecture, much stress will be placed on the fact that hypnosis is a learned activity, one that requires some practice and concentration. The hypnotist will usually add that creative, imaginative, and intelligent people make particularly good subjects, and that not everyone can be hypnotized. Even skeptics in the audience will probably be misdirected by this spiel. They would assume that the hypnotist is merely trying to entice recalcitrant volunteers by flattering potential prospects "smart enough to take advantage of this opportunity"—a classic sales technique—or merely attempting to justify what seems like a superficial undertaking. If this part of the lecture increases the number of audience volunteers, it's fine and dandy. But the main purpose of the hypnotist's stress on the wonderfulness of highly susceptible people is to put the ego of everyone who walks onto the stage on the line. If a person fails to "go under" and "perform," it will reflect badly on them—they are "uncreative" and "rigid" compared to the good subjects. The hypnotist's lecture is also designed to stifle skeptics and would-be hecklers, particularly skeptics and would-be hecklers who are planning on volunteering expressly to prove to their buddies that hypnosis is all fake. If the hypnotist can convince an otherwise cynical person that he or she is inferior if unable to be hypnotized and thus might be looked down upon by the audience, the hypnotist is put back in control.

5. *The hypnotist makes asides to the subjects while onstage, conversations to which the audience are not privy.* O. McGill, whose 1947 *Encyclopedia of Stage Hypnotism* and subsequent *Professional Stage Hypnosis* are classics in the field, recommended that the hypnotist misdirect by turning his or her back to the audience, pretending to do some technical work in inducing hypnosis, and then rather aggressively ask the subjects, out of hearing

range of the audience, to help him fool the audience. According to McGill,

> Ninety-nine times out of a hundred—if the performer has any ability whatsoever, he can "stage whisper" his instructions to the subject, and receive full cooperation. With his back to the audience, the hypnotist is in a perfect position to whisper instructions and requests to the subject in a low voice.

This type of blatant tampering with subjects is rare today, and McGill eventually modified his stance slightly. But he remained insistent that a skilled performer could get a subject to act on instructions even if the subject was not actually in a trance. If a hypnotic trick is misfiring, McGill suggests whispering to the subject and asking the subject to follow the instruction even if he or she doesn't "feel it." If done properly, the audience never knows what is transpiring:

> These intimate asides spoken quietly and personally to the subject are important to your stage handling of the hypnotism entertainment. The audience hears only the major portions of your comments which describe and explain each experiment, but the subject receives full benefit of your confidence which makes him feel responsible to concentrate well and respond successfully to each test. Further, such handling increases the direct influence of your suggestions.
>   . . . This principle of basically conducting two shows at the same time, one for the audience and the other for the subject or subjects is an important factor . . . in your successful staging of the hypnosis show.

All of these pressures upon subjects, some of which are self-inflicted, combine to produce the behavior that the hypnotist desires and can even lead to what hypnotists call simulation.

When subjects are simulating, they are not in a hypnotic trance but act as if they were. Several scientific experiments have indicated that responsive subjects will do what they think the hypnotist wants them to do, not necessarily what they are feeling or imagining at that particular point in time.

A friend, let's call him Dave, told us the story of his first brush with hypnosis. The hypnotist, when first inducing hypnosis, told him, "You feel your arm rising from the armrests of the chair." Dave did not feel his hand rising. He wondered whether the hypnotist's instructions meant that he should try to raise his arm or whether it meant that he would feel his arm move involuntarily. Dave still did not feel his arm rising. Finally, he *told* the hypnotist he didn't feel his arm rising. "Lift your hand off the armrest, and then you'll feel your arm rising," instructed the hypnotist. Dave willfully lifted his hand and did feel his arm rising.

Was Dave hypnotized? The hypnotist said he was. The phobia for which he sought treatment was eliminated through hypnosis. Yet he couldn't perform a simple task like lifting his arm involuntarily. But what if Dave's first hypnotic session had taken place onstage, with hundreds of people in the audience? Would Dave have felt confident enough to say, "I don't feel my arm rising"? Probably not. He most likely would have simulated a hypnotic state, lifted his arm, and succumbed to the pressures of the hypnotist.

So What's This Hypnosis Stuff All About?

Hypnosis has been mystified to the point of ridiculousness. Most reputable hypnotists will admit that it is a form of concentration. Many psychologists see hypnosis as a form of unconscious role-playing, with the hypnotist supplying the scenario. Some experts see hypnosis as little else but the expansion of a subject's ego to include the hypnotist's consciousness. Stage hypnotists understand all of these principles and always see to it that subjects are deprived of any extraneous sensory input. The hypnotist wants subjects attuned only to the hypnotist.

One researcher compared the hypnotist-subject relationship to that of a mother and baby. Just as a baby sees the mother's breast as a part of his or her own body, so subjects see their hypnotic experiences as coming from within themselves rather than through the instructions of the hypnotist. Hypnosis melts the ego, according to Lawrence S. Kubic and Sydney Margolin:

> It is this dissolution of ego boundaries that gives the hypnotist his apparent "power"; because his "commands" do not operate as something reaching the subject from the outside, demanding submissiveness. To the subject they are his own thoughts and goals, a part of himself.

This view of hypnotist as Machiavellian manipulator, written in 1944, would be disputed by many today. The current thinking is that otherwise healthy individuals cannot be persuaded to do anything against their will while under hypnosis, including acting like a fool in front of an audience.

But these scary implications of hypnosis are virtually irrelevant to the stage hypnotist, who cares much more about a smash show than whether subjects were in a deep trance. As Theodore X. Barber, a tireless demystifier of hypnosis, states, "The successful hypnotic entertainer is actually not interested whether or not the subjects are really hypnotized. He is interested in his ability to *con* his subjects into a pseudo performance that appears as hypnotism—to get laughs and to entertain his audience."

For the dirtiest little secret of stage hypnotists is their most potent weapon: *It is not necessary to hypnotize subjects in order to get them to act as if they were hypnotized.* Barber and various other partners have devoted a great deal of time to experimenting with this subject. Barber, for example, will tell subjects, while fully awake, that a brief period of time will seem like a long time. When later asked to approximate that period of time, those subjects planted with the suggestion reported that the time period was much longer than a control group. Barber was able to induce extremely strong hypnoticlike reactions in fully awake subjects. Eight percent of his subjects would immediately respond to at least seven out of eight maneuvers involving physical reactions (e.g., "You can't stand up—try it," or "You are very thirsty . . . dry and thirsty"). Significantly more subjects would respond to one or two commands.

Kenneth Bowers, in his book *Hypnosis for the Seriously Curious*, corroborates that many people "under hypnosis" will respond to similar simple commands, particularly the old favorite "your arm is getting heavy." He claims that 20 percent "can respond satisfactorily to a suggestion that they are unable to smell a bottle of household ammonia." In a random group, someone will always be responsive to almost all suggestions, no matter how difficult, while others will be unresponsive to all suggestions, regardless how simple. The stage hypnotist, of course, must find those responsive subjects.

Barber and collaborator William Meeker theorize in their excellent article "Toward an Explanation of Stage Hypnosis" that the very fact that stage hypnotists define their act as hypnotic rather than as mere suggestion increases their chances of success. Barber and Calverley ran an experiment in which they told one group they would be part of a hypnotic experiment and the other that they were being tested for the ability to imagine. After that point, the two groups were treated identically. They were then tested on a scale of suggestibility that Barber had developed. The subjects in the "hypnosis experiment" proved to be "significantly more responsive to the test suggestions than those told they were participating in an imagination experiment." Why were they more responsive?

Postexperimental interviews with subjects suggest the following tentative answer: When subjects are told that they are participating in a hypnosis experiment, they typically construe this as implying that (a) they are in an unusual situation in which high response to suggestions and commands is desired and expected, and (b) if they actively resist or try *not* to carry out those things suggested they will be considered as poor or uncooperative subjects, the hypnotist will be disappointed, and the purpose of the experiment will be negated. . . .

Since the introduction of one word, hypnosis, into the experimental situation raises subject's responsiveness to suggestions, we can expect a high level of suggestibility in the stage situation which is inevitably defined as hypnosis and which includes a performer who has been widely advertised as a highly effective hypnotist.

Precisely. There *is* such a thing as hypnosis. It has been used effectively as an anesthetic for patients allergic to medicinal alternatives. It has helped countless people regain concentration, dissolve phobias, lose weight, and delve into their past.

But because it is "only" a form of meditation and concentration, the canny stage hypnotist has an advantage over other magicians. Hypnotists have two chances to execute their tricks. If hypnosis fails, then the entertainer's knowledge of human nature, which we have sketched only in bare outline, can be used to induce subjects to act identically to those who are put in a trance. The audience doesn't know the difference. And most of the time, the subject doesn't know the difference either.

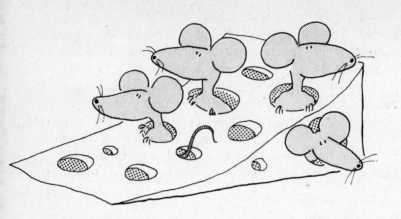

What causes the holes in Swiss cheese?

The cheese industry prefers to call these openings eyes rather than holes. The eyes are created by expanding gases that are emitted by a bacterium known as the eye former. The eye former is introduced during the early stages of Swiss cheese production. The bacterium forms the holes, helps ripen the cheese, and lends Swiss cheese its distinctive flavor.

The eyes, then, are not there for cosmetic reasons. Still, some domestic "Swiss" cheesemakers mechanically "add" holes to already formed cheese produced without the eye-former bacterium. This shortcut is what robs some domestic varieties of the mature flavor of genuine Swiss cheese.

How do they decide which category to put the "Mystery 7" under on *The $25,000 Pyramid*?

Randomly.

How was the order of our alphabet determined? Is there any particular reason why *A* comes before *B* or that *Z* is the last letter of the alphabet?

This is another Imponderable without a tidy answer, and it is necessary to delve into some pretty unpleasant subjects, like ancient history, in order to give it a good crack. We must also admit that this is the condensed, *Classics Comics* version of this story—we aren't even going to bother examining the note-worthy Etruscans, for example. But in order to explain why our alphabet is in the order it is, we have to explore at least five different cultures.

The Egyptians

The Egyptians were writing thousands of years before the birth of Christ. This was the civilization that figured out it might be easier to write on papyrus, with a reed pen, than to carve on stone. Although the Egyptians never created a proper alphabet, their hieroglyphics evolved considerably during the height of their ancient civilization. At one point, they used over 400 different hieroglyphs, but their written language became more and more streamlined as it went through five distinct stages.

1.   Hieroglyphs as pictures of things: A hieroglyph of a horse meant *horse*. This necessitated a separate hieroglyph for every word and promulgated a written language based on things rather than abstract concepts.

2.   Idea pictures: A picture of a leg not only could mean *leg*, but also ideas associated with legs, such as *run* or *fast*.

3.   Sound pictures: One symbol was now used to describe a sound that existed in words of the spoken language rather than as a graphic depiction of the word signified.

4.   Syllable pictures: One symbol represented a syllable of a word. One hieroglyph was now able to appear in many unrelated words that happened to have one syllable in common.

5.   Letter sounds: One symbol now took the place of one letter in a word. With the use of letter sounds, syllable and sound pictures were rendered obsolete, since letter sounds were so much more flexible, even if they necessitated more hieroglyphs to create a word. At first there were hundreds of letter sounds, but as the Egyptians learned how to combine letter sounds, they eliminated many redundancies. Eventually, they reduced the number of letter sounds to twenty-five.

An alphabet is a fixed system of written signs, each of which, in theory, stands for one spoken sound. In an efficient alphabet, all the spoken words of a language should be able to be expressed by rearranging these letters. At the point when the Egyptians developed letter sounds, they were close to inventing an alphabet as we know it. Even though their letter sounds gave them the means to write *horse* by sounding out the phonics of the word rather than illustrating its meaning, the Egyptians clung to the first three types of hieroglyphs, never able to figure out why the best way to express *horse* wasn't to draw a picture of a horse.

The Ugarits

Although the Phoenicians are widely hailed as the inventors of the alphabet, it is now conceded that the first ABC's were in the city of Ugarit, in northwest Syria. A German scholar, Hans Bauer, found tablets that have Ugaritic letters displayed opposite a column of known Babylonian syllabic signs, proving that the Ugarits consciously ordered their alphabet. It is unclear whether this tablet was used as an instructional primer. Although the phonetics of the Ugaritic alphabet were identical to the Phoenician symbols, the actual script was different from the later Phoenician alphabet and from the earlier Egyptian and Semitic languages.

The Phoenicians

The Phoenician alphabet was probably developed around the same time as the Ugarits', but the Phoenicians were much more important in the history of language, for they spread their alphabet throughout much of the world. The Phoenicians weren't aesthetic types. They were traders and needed an alphabet not for literature or history (they didn't leave behind any books) but for business—to track inventories, to standardize accounting procedures, and other such mercenary tasks. By 1000 B.C., the Phoenicians were carrying their alphabet with them to most of the major ports of the Mediterranean.

The Phoenicians totally dropped the picture signs of hieroglyphics and kept only the symbols that signified sound. The Phoenician's word *aleph* meant "ox," and the letter *a* was made to look like an ox's head. The ox, the most important farm animal of the time, was the basis for the first letter of most European and Semitic languages, including, later, English.

The Greeks later adapted the Phoenician language to their needs. They took sixteen characters from the Phoenicians, all consonants. It was up to the Phoenician reader to decide where the vowels belonged in a given word. A headline of a Phoenician critic's review of this book, for example, might have read like this:

"*MPNDRBLS*, PRTTY GD BK"

Every spoken language has vowels and consonants, but a remarkable number of ancient written languages did not include vowels in their alphabet. Technically, a consonant, according to *Webster's New World Dictionary*, is a sound produced "by stopping and releasing the air stream (e.g., [our] p, t, k, b, d, g) or stopping it at one point while it escapes at another (m, n, l, r), or forcing it through a loosely closed or very narrow passage (f, v, s, z)." Consonants are formed by the vocal chords with the assistance of the tongue, teeth, or lips. A vowel, on the other hand, is formed simply by the motion of the vocal chords, with no obstruction by the other speech organs. The lack of vowels in the Phoenician alphabet is about all that keeps it from being a modern language.

The Greeks

The Greeks were scavengers, taking their favorite elements from the Semitic and Phoenician languages and synthesizing their own. Around the ninth century B.C., the Greeks added five vowels to what were essentially the Phoenician consonants, and these are the five vowels that English speakers can recognize, not only because the names of each vowel start with the five letters that are our vowels, but also because the names of all but the "o" vowel have become household words themselves: alpha, epsilon, upsilon, iota, and omikron. Alpha became the first letter in the Greek language.

Sharp-eyed readers might ask how, if the Phoenicians didn't have vowels, their *aleph* metamorphosed into the Greeks' *alpha*. Actually, the *alpha* was taken from the Hebrew language rather than Phoenician, and its similarity to Phoenician is because *aleph* also means "ox" in Hebrew. The first letters of the Hebrew alphabet are *aleph*, *beth*, *gemel*, *dalth*, which mean "ox," "house," "camel," and "door," respectively. The Greek equivalents are *alpha*, *beta*, *gamma*, and *delta*.

The driving element in adaptation of written languages is whether the old language can express the sounds already verbally expressed in the adoptive country. The Greeks needed letters to express vowel sounds that already existed in their spoken language. The Phoenician alphabet did not have them, and although the Hebrew language did have vowel sounds, they were used erratically and sporadically. But the Hebrews did have some consonants that used sounds the Greeks did not have. This was the case with the first letter of the alphabet. In Hebrew, the *aleph* was a soft breathy sound that had no phonic equivalent in the Greek language. The Greeks took such "useless" consonants from the Hebrews and converted them to vowels in the Greek languages. Thus, the Greek vowels were Hebrew in origin and the consonants Phoenician.

By adding a few consonants of their own, the Greeks ended up with a 24-letter alphabet. They had no equivalent of our *c* or *v*, and some of their letters stood for sounds different from their modern equivalents. Their *p*, for example, sounded like our *r*. Still, their alphabetical order is roughly the same as ours today, with several notable exceptions, including the fact that their *z* was the sixth, rather than the last, letter of their alphabet.

The Romans

The Romans were once ruled by the Etruscans, who used the Greek alphabet. Before their decline, Romans adopted the Greek alphabet and then began to make changes. The Romans established the current alphabetical order used by English-speaking countries, but their alphabet contained only 23 letters. *J*, *u*, and *w*, were introduced well after the birth of Christ.

The letter *j* was originally used as a variant of the vowel *i*. Until the seventeenth century, Caesar's name would have been spelled Iulius. The *w* was expressed in Anglo-Saxon by the notation *uu* or *u* until about A.D. 900. The *u* itself was used as a variant of the letter *v*. It wasn't until the eighteenth century that the letter *u* was used exclusively as a vowel.

Why did the Romans rearrange the order of the Greek alphabet? There were various reasons for the changes, perhaps none as interesting as why *z* got dumped at the end of the alphabet. At first, the Romans dropped the Greeks' sixth letter altogether, figuring it was unnecessary. When Rome conquered Greece in the first century B.C., the Romans decided they needed the letter back again, primarily in order to transliterate Greek words into Latin. By this time, however, the Romans had formalized their alphabet, and the *z*, having lost its rightful place in line, got sent to the back of the bus. Other Romance languages haven't seen the need to reassert *z*'s original position.

Clearly, the placement of letters in the alphabet was essentially an arbitrary one. It would make more sense, probably, to have all of the vowels lumped together separately at the beginning or end of the order. Learning alphabetical order doesn't matter much in helping to master English. Would we read or spell any less effectively if we learned the alphabet in reverse order? Yet the Ugarit tablets indicate that the alphabet was taught in alphabetical order, and linguists have found in most cultures that lists of alphabets invariably were written in the same order, despite the fact that unlike numerical order, the order of letters has no intrinsic meaning.

It is the utter serendipity of our alphabetical order that makes the explanation to this Imponderable so disarming. Would anyone guess that *a* comes before *b* because, for an ancient Semitic culture, "ox" came before "house?"

Why does unsweetened canned grapefruit taste sweeter than fresh grapefruit?

Paranoid types suspect food packagers of sneaking sugar into cans and jars of this most austere of fruits. Canned grapefruit simply tastes sweeter than fresh.

Paranoids, relax. The only reason canned grapefruit is sweeter (or unsweetened grapefruit juice is so palatable) is because the fruits are picked from the trees later than those you would find at the produce section of your grocery store. Obviously, if fruits are picked while they are at their peak of flavor, they would be spoiled by the time many consumers had the chance to buy them. Supermarkets avoid this problem by selling grapefruit before it has reached maturity. Spoilage is the bane of produce marketers—after all, what can a grocery store do with overripe bananas but dump them in a box, sell them for 25 percent of the regular price, and hope that someone wants to make banana bread?—so canners have an advantage over their fresh fruit competitors. Canners are able to process the fruits immediately after they are picked.

A few canned grapefruit suppliers sweeten their grapefruit by including the juices of other fruits, but this tinkering isn't necessary. Fresh grapefruit, eaten at the peak of ripeness, belies its image as bitter and astringent.

How can cats see in the dark?

They can't. No animal possesses the ability to give light. Stick a cat in a dark cave and it will not be able to tell the difference between a stalactite and a stalagmite. Cats are no more able to see in total darkness than humans.

What cats *are* able to do is to reflect very faint light rays. When they appear to glow in the dark, cats' eyes are reflecting what little natural or artificial light is available in a seemingly dark environment.

The structure that allows cats to reflect light is called the *tapetum lucidum*, an iridescent layer of cells around the optic nerve. The *tapetum lucidum* is part of the membrane between the cat's retina and the outer covering of the pupils. When light is cast upon the *tapetum lucidum*, it is reflected back as if the cat's eyes were mirrors. All cats, including jungle varieties like lions, jaguars, and tigers, have the capability of reflecting light rays even though they cannot see in complete darkness.

Cats are fundamentally nocturnal animals. Even though they cannot see in total darkness, cats' physiology is designed for them to see much better at night than their bumbling owners. Cats' eyes are positioned more forward on their heads than ours, allowing them a wide field of vision, especially since each eye is capable of overlapping the other's sight. With this binocular vision, cats have two chances to spot a scampering mouse. Cats' pupils are very sensitive to ultraviolet light, enabling them to see things humans cannot. The visual purple in the epithelial rods of their retina allows cats to catch ultraviolet rays.

Less well known is the fact that by comparison to humans, cats are veritable Mr. Magoos during the day. Their slit pupils, which make them look so ominous in Stephen King movies, protect their retinas from the sun. The slits are composed of two hoods that function as curtains, which can be drawn to let in the sun or pulled together to block the sun out. When pulled together, as they often are during the day, a cat's vision is extremely narrow.

So cats' vision is highly overrated. They can't see well in bright sun, not at all in complete darkness. They can outsee humans only in moderate or dark light—like in a house, where they spend all of their time.

How do they get white wine from black grapes?

Most people are surprised when they learn that white wine is often made from black grapes. How can this be? First of all, black grapes aren't exactly the color of 8-balls. Although some black grapes are bluish black, many are really deep red. But this merely begs the question and adds a new one: How do they get white wine out of deep-red grapes?

The answer is simple, and you can demonstrate it to yourself. Go into your refrigerator and pull out the darkest grape you find. Or if you are so inclined, run without delay to your friendly produce merchant and purchase a few of his or her darkest grapes. Place a grape between your thumb and forefinger. Crush it. After wiping the gook off your shirt, look at the liquid that came out. Pretty puny stuff, huh?

Almost any grape juice will be a white or yellowish color. The only reason your red wine is so deep and vivid is that the color is derived not from the juice of the grape but from the fruit's fermented skin. Without the skin, a white wine can be made from just about any color grape. Even champagne is made partly from black grapes.

Why does full service at most gas stations cost much more in proportion to self-service than it did when first introduced? Why does the price differential between self-service and full service vary between the various grades of gasoline at the same service station?

The answer to this Imponderable can't be understood without realizing the enormous changes that have swept gasoline marketing in the last twenty years. The oil crises (famine and glut) have deterred domestic oil companies from spending money on new drilling. The trend toward streamlining operations continues unabated.

The major oil companies have stopped building new stations. In 1974, the average major oil company constructed only 8 new service stations in the entire country and deactivated 295. In 1982, the large oil companies constructed an average of 9 new stations and deactivated 98.

The largest selling item at most service stations—unleaded gasoline—wasn't even available twenty years ago. One of the previous best-selling gasolines—leaded premium—is unavailable at most service stations outside of the West Coast.

The configuration of most gas stations has changed radically in the last twenty years. While most stations offered only full service two decades ago, only about 25 percent of today's gas stations provide full service exclusively. More than a third offer a split-island choice, and self-service-only stations and self-service convenience stores comprise about another third of all gasoline sales in the United States. As recently as ten years ago, almost half of all service stations were full service only.

There are many reasons for these radical changes, and all of them contribute to why the prices at the pump often seem irrational. Here are four of the most important factors in explaining the changing economic picture at the service station.

Legal

A federal law, the Petroleum Marketing Practices Act, dictates that the large oil companies can set only the wholesale price of gasoline. Mobil or Shell acts as a distributor of gasoline, but the service station dealers retain total autonomy as price-setters. The wholesale prices of the major oil companies may vary considerably, depending upon the costs of transporting gas to that location and upon the local supply-and-demand situation.

Dealers' Costs

The cost of gasoline to the dealer is called the rack price. This is the wholesale price to the dealer if he or she picks up the gasoline and transports it back to the station. The distributors will deliver the gasoline for a price, which can be substantial considering that the average service station possesses three underground tanks that hold a total of approximately 25,000 gallons of gasoline.

At the time of this writing, the wholesale price for gasoline is not out of proportion to the retail one. Two different oil companies estimated that their average wholesale rack prices were 73 cents for regular leaded, 75 cents for unleaded, and 80 cents for super unleaded. Regular leaded will almost always cost less than unleaded, and unleaded, of course, will always cost less than premium leaded.

The biggest variable in service station costs is labor. Even all self-service stations must hire cashiers. Service bays will increase revenue (there are a lot of bucks in mechanical work) but will also greatly increase inventory costs and space requirements. And unlike the guy who pumps gas, mechanics cannot be found who will work for minimum wage. If the owner can find gas jockeys for minimum wage and can charge enough extra for full service, more profits can be realized. But splitting service bays can mean long lines and impatient customers when the self-service island is clogged and the full-service island is empty. And worse, full-service customers are less than thrilled when employees are busy waiting on other full-service customers and they eye folks getting in and out at the self-service pumps before they even get to the pumps. If the station converts to all self-service, loyal customers might leave in droves to search for full-service establishments. So although it is easy to estimate the costs and economies of self-service *vs.* full service, it is harder to prejudge its impact on the volume of customers and the retention of their loyalty (e.g., "If we convert to self-service, will our old customers simply flock to the service station with the best prices?").

Conservation

Not only are there fewer gas stations than there were ten years ago, there is less gas being sold, despite the increase in the number of cars on the road. From 1975–1982, the total consumption of motor gasoline went down by over 10 percent.

Almost every major accessory that gasoline stations used to sell has been improved in performance and reliability in the last ten years. We have gone from a V8 society to a 6-cylinder one. Newer cars need changes of oil much less often. Tires and batteries last longer.

For the service station, there is only one redeeming statistic in our nation's successful oil conservation drive—because of the attrition in service stations, more gallons of gasoline are now sold *per terminal* than ten years ago. And more gasoline per service station.

Today's service stations are clearly leaner and meaner than they once were. They were forced into focusing on gasoline rather than accessories and repairs because of the threat of . . .

Competition

Until the mid-seventies, conventional service stations sold about one-third of the nation's tires, batteries, and accessories. This share plummeted suddenly in the mid-seventies, and although the drop has leveled off and the dollar value of these sales is more than that of ten years ago, service stations have permanently lost market share and seem not to want to fight to reclaim it. Sales of motor oil, particularly, have sprung a leak, as supermarkets, drug stores, discount stores, and other retailers went after the market aggressively. Specialized tire stores effectively launched advertising blitzes focusing on "name-brand" tires, hitting service stations on their vulnerable spot (most customers don't trust a gas station tire as much as a "household name" like Goodyear or Firestone). The one area in which service stations seem to be holding their own is in batteries, lubrication, and exhaust-system replacement, areas where do-it-yourselfers might fear to tread.

More dangerous competition than the specialty and discount stores are convenience stores, since they threaten service stations' sales of their most important product—gasoline. Southland's Seven-Eleven stores, for example, must be reckoned with as a major force in gasoline marketing. There are now over 50,000 convenience stores in the United States and about half of them sell gasoline. In 1982, according to the

American Petroleum Institute, the average convenience store had 5 hoses and pumped over 38,000 gallons per month.

Convenience-store owners find gas sales synergistic. Gas patrons, on the way to pay their bills, make impulse food purchases. Others, out for a six-pack, figure they might as well kill two birds with one stone, and gas up. In almost every way, the convenience store has a cost advantage over the conventional service station. The construction of a convenience store costs one-half that of service stations, and gas pumps are usually located in areas of the parking lot that were previously not utilized. The salary of store managers at convenience stores is much lower than that of gas station dealers.

Pricing Strategy for Conventional Service Stations

Conventional service stations are pulled in two directions. On the one hand, since an increasing percentage of their volume lies in gasoline sales, they must be assured that their profit margin on gas is high. On the other hand, faced with the low-balling price competition of convenience stores, which are invariably self-service, the ultra-low-price strategy of some major oil companies (such as Arco, which has reduced prices by eliminating credit cards) and less expensive independent service stations ("generic" gas), service stations must compete in price to survive. Twenty years ago, customers were not so price-sensitive, largely because the service station was not viewed solely as a gas dispenser, but also as a one-stop automobile clinic, where it paid to establish rapport with the people who maintained and fixed cars. Now, when all gas stations provide essentially the same products and cannot offer service as a distinguishing benefit, price becomes the only significant difference among service stations for many consumers.

Most states require service stations to prominently post the prices for at least some of their types of gasoline. Even in states without this legal requirement, low-ballers want to announce their prices, and "name-brand" service stations are viewed with suspicion if they fail to post prices.

To entice customers into their stations, gasoline marketers have adopted one of the oldest merchandising techniques: the loss leader. In most stations, regular leaded gasoline is sold at a disproportionately low cost, so that the station may post a low price prominently on its signs. Most dealers work on what is called the pooled margin concept. Rather than have each type of gasoline carry its own weight in profit margin, the goal is to have the mix of the three grades and the self-service and full-service options yield together the desired profit margin. If no profit is made on the self-service leaded regular (the perfect type of gas to be the loss leader, since it is the cheapest and has a rapidly declining share of sales), the loss can be made up for by artificially increasing the price of, say, full-service unleaded premium.

Some states actually have laws prohibiting the sale of any grade of gasoline at a loss, in order to prevent unfair price manipulation (and the forcing out of small dealers by larger oil companies). Despite all the fancy strategies, the name of the game in gasoline pricing is: What will the market bear?

At the time this book is being written, this might be the price structure of a typical conventional service station:

	Self-service	Full Service
Regular	$1.12	$1.37
Unleaded	1.24	1.49
Premium Unleaded	1.42	1.69

When self-service was first introduced, the pricing gap between self-service and full service was considerably less than it is today. At the time, most dealers didn't realize how many customers would switch to self-service and how many employees could therefore be laid off. But the original 3–5 cents that was charged as a premium did not pay for the use of even a minimum-wage employee in most circumstances, and competition from totally self-service stations forced most dealers to cut prices. Profit margins were wrecked.

Not only did most station owners find that raising the prices significantly did not deter most full-service customers, they found that full-service customers in general were not price resistant. This insight led to all sorts of pricing inequities, including the Imponderable that originally inspired this question: Why must the premium unleaded customer, in our hypothetical gas station above, pay 5 cents more for the privilege of receiving full service than his unleaded or regular counterpart? Surely it doesn't cost more to pump premium than regular!

These strange pricing gaps—and they are very common—exist because dealers can post them without customer resistance. The extra nickel per gallon that the dealer picks up on premium unleaded customers allows him to advertise the regular leaded for a few cents less, perhaps dragging in a few more hapless customers.

Why does a thumbs-up gesture mean "okay"?

Any Roman gladiator movie worth its salt includes the obligatory Colosseum combat scene wherein the sated, corpulent sovereign seals the fate of a beaten warrior by giving the thumbs-down signal. The crowd, with their thumbs, advises the king; in movies, at least, hanging juries were the rule, though they may have been prejudiced by the foreboding music on the soundtrack.

Most people assume that the modern-day thumbs-up signal originated from this Roman affirmative meaning. Not so. In his book *Gestures*, Desmond Morris explains that Romans signified their approval of a beaten warrior not by signaling thumbs up but by covering their thumbs. When the crowd wanted the victorious gladiator to finish off the loser, they extended their thumbs, which Morris theorizes mimicked the act of stabbing the beaten man.

If Rome was the birthplace of the thumbs-up signal, we would expect the gesture to be popular there today, and yet Italy (followed by Greece) was found to be the country in Europe where this meaning is least signified. In many parts of southern Italy and Greece, the thumbs-up gesture is a sexual insult rather than a sign of approval. It is likely that the thumbs-up gesture started somewhere else.

If the early Roman derivation has been debunked, why do we use a thumbs-up signal to indicate "okay"? The historical evidence, as with most gestures, is murky and contradictory. Morris and other authorities believe the predominant reason is that Western culture tends to associate upward movements with positive, optimistic feelings and downward movements with negative, pessimistic emotions. Obviously, any finger pointed upward is heaven-bound. In the 1970s, the gesture of a forefinger extended upward became a symbol of fundamentalist Christians. The solitary forefinger not only indicated "one God" and "one way," but where God resided and where the good Christian could someday reside.

The thumb might have been selected as the raised digit because it is the most easily isolated finger. Try raising your finger and withdrawing your other fingers, and you'll realize why the thumb was a more natural choice.

Why can't you buy Hellmann's mayonnaise in the West? Why can't you buy Best Foods mayonnaise in the East?

If you live in or west of Montana, Wyoming, Colorado, or New Mexico, chances are you buy Best Foods mayonnaise. If you live in or east of North Dakota, South Dakota, Nebraska, Kansas, Oklahoma, or Texas, you probably buy Hellmann's mayonnaise. Both brands are dominant market leaders where they are sold, but except for the El Paso area of Texas, their distribution does not overlap at all.

Why don't the two brands compete against each other? It's a long story.

First there was Hellmann's—actually, *a* Hellmann, a German immigrant who opened a delicatessen in New York City in 1905. His wife created a mayonnaise that Richard Hellmann used in his own store. In response to customers' requests, he began selling the mayonnaise "to go"—for 10 cents.

Hellmann originally sold two slightly different formulas. His

---

most popular formulation's jar was adorned with a blue ribbon, which still is displayed on Hellmann's jars today. Hellmann's cottage industry soon turned into big business. Once he started prepackaging jars, in 1912, distribution spread rapidly. He bought a truck to deliver jars, and soon he had a fleet. In 1913, the first Hellmann's plant was built on Long Island, and the mass-market mayonnaise business blossomed in America.

Hellmann's success caught the eye of giant General Foods Corporation. In 1927, it purchased the Hellmann name and facilities. General Foods extended the distribution of Hellmann's beyond the East Coast, since competitors were sprouting throughout the country.

The most successful of Hellmann's counterparts was a West Coast mayonnaise called Best Foods. Best Foods wanted to seek eastern markets, and an inevitable clash was averted when Best Foods took over the Hellmann's brand in 1932, thus gaining the lion's share of the real mayonnaise market in the United States. By the time of the merger, both brands were so firmly entrenched in their areas, and had such a dominant market share, that it was decided not to change either name. Best Foods later became a division of CPC International, Inc., and CPC decided not to rock the boat either. West of the Rockies, Best Foods is the number one mayonnaise; East of the Rockies, Hellmann's is king.

CPC is a canny food marketer. For the most part, its brands are in extremely unsexy categories—syrups, oils, pasta, bread, peanut butter, and dyes. CPC makes no attempt to trumpet its own name in advertisements, as Beatrice and General Foods do, yet virtually all of its brands—Hellmann's/Best Foods, Karo, Mueller's, Niagara, Rit, Skippy, Thomas', and Knorr—are market leaders in the U.S. and abroad. CPC believes that brand identification is key—that consumers are loyal to Tropicana orange juice, not to Beatrice—so CPC is willing to suffer inconveniences in order not to interfere with allegiance to its popular brands.

One obvious problem with manufacturing two mayonnaises is advertising. Although the advertising campaigns for Best Foods and Hellmann's are identical, they must not only make two separate commercials for each product, but then are incapable of running national television or radio spots or national magazine ads: CPC is forced to use "split runs," losing some economy of scale. Extra label and packaging costs are incurred by retaining the separate identities of the two brands.

The obvious questions remain: When Best Foods took over Hellmann's, which recipe was used? And are the two mayonnaises now identical? When Best Foods took over, the new Hellmann's labels assured the consumer that the old Hellmann's formula was being maintained. The mayonnaises were quite similar anyway, with Hellmann's having a slightly spicier flavor. Even today, the products are not identical. Although they are processed in the same way, their seasonings are a little different, and Best Foods mayonnaise is slightly more tangy. Both labels include the information that the mayonnaise is known by the other name on the other side of the Rockies, but representatives from CPC informed *Imponderables* that most consumers don't know about the Hellmann's/Best Foods connection and tend to complain to grocers when they can't find their favorite mayonnaise in a new location.

CPC was faced with another marketing problem when deciding what to call its mayonnaise in foreign countries. CPC's mayonnaise is sold in about twenty countries overseas, with Hellmann's the most common name. But Best Foods is used too, and the same mayonnaise is sold as Fruco in much of South America and as Chirat in Switzerland.

Considering the potential confusion, and the fact that no other food product is a market leader under two different names, it is a bit surprising that CPC has never considered aligning the two brands. CPC insists it is happy with the arrangement, and when you consider that, divided, it has conquered such supermarket heavyweights as Kraft, it's difficult to argue with CPC's logic.

Not all companies are content to be anonymous to the consumer or to see conflicting brand names diminish the strength of their identity. In late 1984, Nissan Motor Corporation bit the bullet, and the world's fourth-largest automobile manufacturer announced that there would be no more Datsun. All of its products would now be manufactured under the parent company's name—Nissan.

Nissan sells vehicles in more than 150 countries and manufactures or assembles cars and trucks in 21 countries, including the United States. Nissan felt it was important to ensure worldwide continuity of the Nissan name. The cost of the project: roughly $30 million, including installation of new signs, removal of old signs, and design changes. Nissan then undertook a massive advertising campaign to emphasize that Datsun had become Nissan.

In a case where a brand has a poor image, name changes are often tried (e.g., Sambo's coffee shops), usually without much success. It is unusual for a successful name to be retired, but as the president of Nissan Motor Corporation in the U.S. stated, "Two names can be confusing, especially as Nissan's global operations grow."

Nissan figures that if you like the Sentra, you might be prone to buy a Nissan if you were in the market for a truck. CPC figures that even if you like Hellmann's or Best Foods mayonnaise, you won't automatically buy its Mazola instead of Wesson. Nissan's strategy is to build corporate loyalty; CPC's strategy is to build brand loyalty.

Why are copyright dates on movies and television shows written in Roman numerals?

This is not the kind of question movie studios want to answer on the record. About the only reason anyone could come up with to answer this Imponderable is the obvious one—they express the release date in Roman numerals in order to make it more difficult for viewers to determine exactly how old the show is. It is hard enough to spot the release date printed in Arabic numbers during a fast credit crawl.

Although studio representatives were not unwilling to so speculate off the record, none of them knew this "deception theory" to be a fact. It may be just as likely that copyright dates are in Roman numerals simply because they've always been that way: Never discount inertia as an explanation for *any* phenomenon.

There are many new avenues for international distribution of movies and television shows, notably cable television, home video, and videodisc. With each new "window" of distribution, some time elapses. A hit movie might show up on cable television six months after its theatrical release, and then on videotape and videodisc a year later. But non-hit movies can have a more erratic distribution time frame—B-movies like horror movies or kung-fu flicks might not even hit theaters, let alone home video, until many years after they are shot. And the movie may not be released in foreign markets until even later. There are more reasons than ever for concealing the true release date of movies (writers have long made it a policy not to put dates on screenplays they send out for consideration—the older the date, the staler the script somehow seems to the reader).

Ironically, though, more and more Arabic numbers are popping up on release dates, particularly in television. ABC- and NBC-produced shows now use Arabic numbers, and some movie studios use Arabic numbers, although the policy is inconsistent. W. Drew Kastner, a lawyer for NBC, indicated that the network has no reason to make it difficult for viewers to know exactly when a show was taped or filmed.

Is there any practical reason for the copyright date in the first place? Although ideas cannot be copyrighted, the expression of such ideas is protected. By inserting the copyright date, movies are automatically protected by the Universal Copyright Convention, which means that if there is a copyright date listed, it will be protected internationally, without the need for costly legal paperwork in each locality the film is exhibited. Under the old copyright law in the United States, the term of the copyright was 28 years from the date of publication. But under the current law, effective January 1, 1978, the copyright extends to the life of the author plus 50 years, or 100 years after creation, or 75 years after publication, whichever is less. With the advent of home video, the copyright on a film is more valuable than ever. It isn't important, or even desirable, for you to be able to read the copyright date while watching the movie. But it is important that would-be plagiarists know where they stand.

Why don't penguins in the Antarctic get frostbite on their feet?

The yellow-necked emperor penguin, the largest species of penguin, spends its entire life resting on snow or swimming in water at a below-freezing temperature. A penguin's dense feathers obviously provide insulation and protection from the cold, but how can it withstand the cold on its feet, when humans won't put their limbs into the ocean when water is in the sixties?

Penguins' feet are remarkable creations. They are set back much farther than on other birds, so that penguins walk upright, but this conformation's main attribute is to help them swim. Next to the dolphin, the penguin is the fastest swimmer in the ocean. When swimming, a penguin's foot trails behind in the water, acting as a rudder and a brake.

During their hatching season, mother and father alternate diving into the ocean for food. *Encyclopedia Britannica* estimates that the cooling power of the seawater to which they are exposed is the equivalent of a temperature of −4 degrees Fahrenheit with a wind of 70 miles per hour. Add the 10 or 20 M.P.H. speed at which the penguin typically swims, and you have rather uncomfortable conditions. The penguin's skin is protected by a layer of air trapped under its feathers—only the feet directly touch the water.

When the penguin finds food, returns to the mate, sits on the chick, and watches the mate leave to find more food, it has gone from the frigid water to standing directly on snow that is, needless to say, below freezing temperature. How can the feet withstand such punishment?

Penguins' feet do get very cold. They have been measured at exactly freezing, in fact. If their feet stayed at a warmer temperature, they would lose heat through convection or conduction.

The low temperature is maintained by penguins' unique circulatory system. As arteries carry warm blood toward the toes, penguins have veins right next to them carrying cold blood back in the opposite direction. In effect, the two bloodstreams exchange heat so that the circulation level can remain low enough to conserve heat and just high enough to prevent tissue damage and frostbite. Penguins' feet have very few muscles. Instead, their feet possess a vast network of tendons, which do not become as painful as muscles when cold.

Of course, there is another explanation for why penguins don't exhibit foot pains. They are not crybabies, and they are tougher than humans.

Why haven't the fast-food chains been able to create a successful dessert?

Fast food. Junk food. The two seem to fit together like hand and glove. But what do we see? Ads for Wendy's introducing a line of low-calorie foods. Burger King installs salad bars.

For a long time fast-food chains have developed desserts, and for a long time they have failed. In 1984, desserts constituted only 4 percent of total fast-food revenue. Isn't the customer who wolfs down a cheeseburger and fries a prime candidate for dessert? Aren't kids, devotees of fast food, the largest consumers of dessert?

It turns out that there are at least ten reasons that conspire together to rob the fast-food chains of the profit they feel they deserve from desserts—and in no particular order, here they are:

1.   Sixty percent of fast-food revenue is collected during breakfast and lunch. Most desserts are bought at dinner or as an after-supper treat.

2.   The big fast-food chains face specialized competition. Ice cream parlors and cookie stores are able to put all of their marketing muscle on their desserts and usually serve a superior product.

3.   Desserts are often an impulse purchase, yet a customer must make a dessert decision when ordering the entree. Unlike cafeterias, which often place desserts right at the front of the line, there is little visual stimulation to entice a customer to buy a dessert in a fast-food establishment.

4.   Even if a customer decides that he or she feels like eating a dessert after consuming the rest of the meal, it is necessary to trudge back into line for the one item. A few fast-food companies have experimented with "hostesses," who come around to tables and ask patrons if they are interested in ordering dessert. Although this response to the "laziness factor" increased dessert sales, it didn't stimulate sales enough to pay for the added labor costs. Desserts in fast-food stores are what restaurateurs call an "add-on" purchase—no one enters McDonald's just to order McDonaldland cookies.

5.   There are technical problems with serving any frozen dessert. McDonald's experimented with Tripple Ripple, a pre-

fab ice cream cone that seemed to defy science by staying hard in perpetuity, even when exposed to heat. McDonald's was clearly trying to develop a product that wouldn't melt if a customer ordered it at the same time as the rest of the meal. Its sequel to Tripple Ripple turned out to be soft ice cream (sundaes)—they might melt, but only into the cup container. A few fast-food chains have experimented with do-it-yourself sundaes, which have often created *Animal House*-type messes. The other problem with sundaes is that the fast-food stores must underprice their product compared to ice cream specialty stores if they want sales.

6.   Customers seem extremely price sensitive about desserts, and fast-food customers, in general, are more concerned with food costs than their sit-down counterparts.

7.   Even the customer who always orders dessert in a sit-down restaurant is less likely to do so at a fast-food store. While a leisurely feast at the local steakhouse might be seen as a treat or reward by the customer, a visit to the local Burger King is more likely to be viewed as a pit stop.

8.   Up to 80 percent of some fast-food chains' sales are in meat-connected items. It is simply a waste of money to spend promotional money on hot apple pie when hamburger sandwiches are their—no pun intended—bread and butter.

9.   Most desserts have a lower profit margin than other menu items. Given a choice, a restaurant would rather promote soft drink sales (as both McDonald's and Burger King have done in the past) than desserts, since the profit margin on a soda can be 90 percent. Since fast food can only be served fast because the restaurant offers a limited menu selection, there is little incentive to add new dessert items, which are a tough sell to begin with and which are lower profit-margin items.

10.   Most fast-food desserts aren't very good. You may argue that most fast-food *hamburgers* aren't very good. And you'd have a point. The difference is that the alternatives to a bad fast-food hamburger are cooking one yourself or going to a full-service restaurant with longer waits, bigger bills, and obligatory

tips. The alternative to a bad dessert is the local ice cream parlor, bakery, or convenience store.

One fast-food manager we spoke to felt strongly that a lack of quality was the main reason the industry has been unable to create a breakthrough dessert. His store introduced homemade chocolate chip cookies, a risky entry since it competed not only with store-bought cookies but also with the chains of "gourmet" cookie stores that are spawning throughout the country. If his cookies were baked on-premise, there was simply not enough turnover of product to justify the labor required to cook them. If he kept cookies hanging around indefinitely, they became hard and lost the homemade texture that was their main selling point. If he tried to reduce the number of cookies they cooked to ensure freshness, then there would be a sudden burst of customers who would become impatient at the thought of waiting for fresh cookies to be made. If the store, in desperation, turned to prepackaged chocolate chip cookies, with a much longer shelf life, they tasted more like airport food than homemade food. Although his just-out-of-the-oven cookies tasted swell, he couldn't coordinate the timing with enough volume to make the dessert cost-efficient.

In short, making desserts is a pain in the neck, and it isn't worth the time of management to kill themselves perfecting products that contribute so little to the bottom line. Most fast-food chains built themselves by stressing one product. A few, like Roy Rogers, have managed to create a more diverse image by producing credible products in several categories (hamburgers, chicken, and roast beef). But none so far has come up with the dessert special enough to bring customers through the door, and this phenomenon is unlikely to change in the near future.

Why is film 8, 16, 35, and 70 millimeters wide?

Eight is half of 16. Thirty-five is half of 70. But 16 isn't half of 35. Is there any logic to how film widths were derived?

Actually, there is . . . sort of. If you've been reading *Imponderables* with the proper diligence, you have already learned that Leica licked the problem of how to make a big picture out of a small negative by using cine film in still cameras. What you might not know is that Thomas Edison and his assistant, W.L.K. Dickson, were developing a motion picture camera at about the same time that George Eastman introduced his first camera.

The film for the first Kodak camera in 1888 used a paper base and a strippable gelatin emulsion. After processing, the emulsion had to be transferred to hardened gelatin "skins" so that the negatives could be printed. A nuisance. But only one year later, Eastman introduced transparent film base, which eliminated the need for "skins."

The film for the first Kodak camera was 2¾ inches wide, or 70 millimeters. Kodak has been manufacturing 70 millimeter film continuously since 1888.

Thomas Edison was excited about Eastman's transparent base, and he obtained this 70 mm film. Edison wanted to use narrower film for his camera and tried using ¼ and ⅓ of Kodak's width. Edison and Dickson finally settled on one-half of Kodak's 70 mm width, or 1⅜ inches. The Eastman Kodak Company informed *Imponderables* that right from the start, Europeans referred to Edison's film as 35 mm, whereas it was often called 1⅜ or "Edison standard" film in the United States.

The smaller sizes of film were introduced later, as a Kodak historian related:

Sixteen millimeter was derived from tests that began before 1916, when it was determined that a picture ⅙ the size of the standard cine frame would produce a satisfactory image. To this image size of 10×7.5 mm, edges of 3 mm were added for the perforations. This also divided evenly into the width of the film base, so that 70 sixteen-millimeter "cuts" could be made across the width of the film coating. Eight millimeter is obviously derived by splitting 16 mm film.

There have been other film widths marketed that have not proven so enduring. Kodak tried splitting 35 mm film with one row of perforations, and in a different format, 35 mm film was split into 16 mm, 21 mm, and 22 mm for the Edison Home Kinetescope.

Although 8 and 16 might not be even divisors of 70, all three of the other standard film widths were developed in direct response to Eastman Kodak's 70 mm camera introduction of 1888.

What is the purpose of a flat toothpick?

Our personal opinion on this subject is strong. The only purpose of a flat toothpick, as far as we can see, is to promote trade for dentists, who must spend a significant minority of their time trying to dig out pieces of splintered wood from the mouths of misguided naifs who prefer flat toothpicks. Perhaps history could explain why flat toothpicks have been inflicted on an innocent world.

Perhaps history *could* tell us, but it seems that the history of the toothpick doesn't seem to inspire many writers or encyclopedia editors. The Forster Mfg. Co., of Wilton, Maine, claims that its founder, Charles Forster, introduced the toothpick in the United States. (Until then, would be flossers had to be content with gold stickpins.) Forster, on a trip to South America, saw natives picking their teeth with slivers of ivory. Forster figured that he could make toothpicks out of wood, specifically the white birch that was plentiful around his home.

There are other rivals, however, for the title of the Christopher Columbus of toothpicks. *Famous First Facts* indicates that as early as February 20, 1872, Silas Noble and James P. Cooley of Granville, Massachusetts, were awarded Manufacturing Machine Patent #123,790 for a machine that made it possible for "a block of wood, with little waste, at one operation, [to] be cut up into toothpicks ready for use."

There seems to be no doubt that the flat toothpick preceded the round toothpick and that the round toothpick was one of the less publicized introductions of the 1904 World's Fair. We asked several toothpick manufacturers what advantages a flat toothpick might have over a round one. Nobody could come up with a single advantage, except that flat toothpicks are cheaper—usually three times cheaper. They are cheaper, of course, because they are flimsier.

The evidence seems to be that folks who use toothpicks primarily for flossing or as an adult pacifier prefer round toothpicks. Forster reported that in sales to individuals, their round toothpicks outpace flats about four to one. But about half of all toothpicks are bought by institutions, who use them primarily for food preparation. Some kitchen tasks can be performed as well by flat toothpicks as by round ones (although those who use the "toothpick test" to determine whether baked goods are done might dispute this), and these institutions figure that since toothpicks are a disposable item, why pay three times as much for a round one as a flat one? What they fail to reckon with, of course, is that a toothpick is supposed to be disposed with *after* it is used, not *while* it is being used.

How can the relative humidity be under 100 percent when it is raining?

Air moves in layers. Often, rain occurs when a higher warm, moist air mass overwhelms a cool, dry air mass at ground level.

Humidity is measured at ground level. When the rain from

the higher layer falls through the dry air layer, the humidity on the surface rises, but need not rise to 100 percent. Conversely, when the moist layer is below the high pressure system, the humidity can reach 100 percent on the surface even if the upper air layer is dry.

Why aren't there national brands of milk? Why aren't there national brands of fresh meat?

Scan through a few issues of *Consumer Reports* and it will become apparent that in blind tests, most people have less than discriminating taste buds. Yet through advertising and marketing, fierce brand loyalties can be developed in products that are virtually indistinguishable. Would you be willing to bet your life savings that you could tell the difference between Smirnoff's and Gordon's vodkas?

Since the retail dairy industry is a huge one, it has always been surprising that there is no national brand to cash in on our nation's proclivity for milk. There are a few brands, such as Sealtest and Borden's, that pop up in different parts of the country, but none that can be universally found in the dairy case the way Kraft cheese or Dannon yogurt can.

Is there any legal or regulatory reason why there can't be a national brand of milk? No.

Can milk be transported across state lines without spoiling? Sure, and to some extent, it is done today, particularly from dairy-rich states (e.g., Wisconsin) to populous states in the South that don't have much of a dairy industry. But this begs the question—a national brand could easily be supplied by a cooperative of farmers from all over the country. The national brand would be a marketing concept, not a specific group of cows located on one huge farm.

Would it be impossible to sustain brand loyalty for milk? As Bruce Snow, of the Dairylea Cooperative Inc., told *Imponderables*, "Milk is recognized by the public as something of a 'commodity' rather than a differentiated product. The public assumes, and mostly with reason, that all milk, regardless of brand name, is about equal in quality." This is a totally logical answer and an important factor in the equation, yet the thought still nags that consumers are willing to spend twice as much for a dishwashing liquid that has a certain smell because they (wrongly) think it will be more effective against dirt, even if it is no more effective than its cheaper competitor. Advertising can create brand consciousness and brand loyalty.

There are two other reasons—economic ones—why large corporations fear to tread in milky waters. The first, and perhaps most important reason that Kraft, for example, would be unlikely to want to compete in the retail dairy case is that profit margins for milk, all along the distribution chain, don't meet their fiscal expectations. According to Snow, the profits for the processor are ". . . very small. Profits are in the neighborhood of 1 to 2 percent. Only by marketing in large volumes of a perishable product used daily do processors survive. The big companies want 15 percent return, not 1 percent." Although Dannon also produces a perishable product, yogurt commands a significantly higher profit margin.

If the problem is profit margin, why don't the mega-corporations advertise heavily and push up the price of milk to fatten profits (this is what happened with orange juice, once thought of as a generic commodity). This scenario would probably fail for the second economic reason that Mr. Snow cites— the milk industry is a highly localized one, where local farmers could undercut the efforts of a national brand.

There are hundreds of tradenames on dairy shelves today. Processors collect local milk from farmers or co-ops and many never haul it more than fifty miles to the retailer's. The regional companies have little advertising or merchandising expenses. Their transportation costs are low, and they have no need to store huge quantities of milk, since they collect milk and stock it in stores on a daily basis. How could a national brand compete in price with the local company, whose overhead is considerably less?

But why not branded meat? Sure, meat shares some of the same problems as milk as a branded item. In particular, consumers are used to treating fresh meat as a commodity. We may patronize one store over another because we like the meat better, but most of us doubt that there could be a uniformity in quality for fresh meat anyhow. If Oscar Mayer supplied fresh meat as well as packaged meat, could it guarantee its beef would always taste the same, always be fresh? Would we believe Oscar Mayer if it did?

The only problem with the "commodity" argument is that it has been refuted by the poultry industry. Holly Farms and Perdue chickens, while not available everywhere, have proved that with advertising, a fresh flesh product could be sold effectively throughout the country and could develop intense brand loyalty. Shoppers are willing to pay a few cents extra a pound for their favorite brand of chicken. If so, why wouldn't they pay more (and buy more) for branded meat?

Unlike the milk industry, the meat industry has made serious attempts to create national brands of meat. After World War II, one meatpacker shipped beef and individual steaks to markets precut in consumer-size packages. The meat was frozen and displayed in the frozen-food case rather than the meat area. Shoppers stayed away in droves. V. Allan Krejci, director of public relations at Geo. A. Hormel and Co. informed *Imponderables* that Hormel, in an effort to develop brand identity, introduced branded boxes and inserts for use in packaging Hormel meats at the supermarket level. But when the grocery store butchers cut the primals into consumer-size packages, "the product lost its identity."

Why didn't Hormel cut the meat itself and send it to the stores in individual packages labeled HORMEL? Hormel (and other companies) know that consumers will reject individually wrapped frozen "fresh" meats, and packaging fresh meat involves one tremendous problem not encountered by other perishable foods such as cheese, yogurt, and even poultry. These other products are kept from spoiling during transport between processor and wholesaler and between wholesaler and retailer by oxygen-proof or vacuum-sealed materials. Meat also uses vapor-proof packaging when it travels between wholesaler and retailer, but it cannot be kept in this "boxed meat" stage when it is displayed, because this beef has a purplish-red color that, according to National Live Stock & Meat Board merchandising specialist Thomas J. Flaherty, is unacceptable to consumers. Consumers desire a bright cherry-red color for beef or a grayish-pink color for pork, and even though the "undesirable" color of boxed meat in no way indicates an inferior product, shoppers will not be moved. Meat can be successfully frozen in the desired colors, but as we have seen, consumers wouldn't buy frozen meat if it came in designer colors.

Without the ability to prepack meat before it hits the retail counter, it is doubtful that nationally branded meats can be successful, and yet supermarkets have an obvious objection to the practice. When fresh meat is cut and trimmed of fat, the natural juices of the meat tend to purge, resulting in a shorter shelf life and less eye appeal. The retailer doesn't want to get stuck with day-old meat that becomes difficult to sell. By having in-store butchers, the store can determine how much meat it can sell in one day while having the luxury of cutting more meat on the spot if there is demand.

These in-store butchers are the other major reason nationally branded meat has not caught on, and they may be the key to why we may see nationally branded meat in the future. Until recently, packinghouse meatcutters have commanded larger salaries than retail butchers. With problems in the meat processing industry, labor unions have lost much of their clout, and store meatcutters now earn more than their packinghouse counterparts. The result? Large meat processing companies can now package fresh meats at a price cost competitive with un-branded supermarket meat.

Not only has the labor picture changed, but so has the packaging picture. New vacuum-packaging technology, including a more durable sealing film, has significantly increased the shelf life of fresh meat. Hormel and Wilson are already offering fresh pork in consumer-size packages. If they are successful in their ventures, national advertising and branded meat may not be far behind. Flaherty predicts that we'll see branded beef in supermarkets by 1990.

We have all seen signs saying ALL MAJOR CREDIT CARDS ACCEPTED. What is a *minor* credit card?

Whichever credit card you happen to be carrying that they won't accept.

Why do dinner knives have rounded edges?

Since the purpose of a knife is to cut, why do dinner knives have rounded edges, necessitating "steak knives" to accomplish serious comestible surgery?

Actually, knives, ancient eating utensils, did have sharpened points until the seventeenth century, when the renowned Cardinal Richelieu changed all that. Richelieu, an eater so finicky and fastidious as to make Felix Unger look like a slob, objected to a houseguest who used the point of his knife as a toothpick. The next day, Richelieu ordered his steward to round the ends of all the cardinal's knives. Richelieu's style of cutlery spread throughout France and most of the Western world. By the nineteenth century, most decent folks had a difficult time spearing their peas with their knives.

Why is California or New York sparkling wine called champagne and Italian or German sparkling wine not called champagne?

Sparkling wine has been available for over two thousand years. Although champagne is now renowned for its bubbles, the first champagnes were still wine; even today, a still champagne is produced in France.

Dom Perignon, generally credited with inventing champagne, was a Benedictine monk who was put in charge of wine making from 1668–1715. Peasants would pay their tithes by contributing wines from the local vineyards. Perignon blended the wines in order to achieve the proper balance of sweetness, texture, and finesse.

Before Dom Perignon, the practice was to cover wine bottles with bits of cloth soaked in olive oil. Perignon used a bark of cork to stop his champagne, enabling it to sparkle longer. By firmly sealing the wine after the first year's fermentation, the carbon dioxide, trapped in the bottle with no place to escape, went down into the wine and acted on the sugar in the fermentation process, creating the effervescent effect prized by champagne lovers. Most champagne makers believe that the smaller the bubbles (or beads, as the in-folk call them), the better the champagne. Most champagne is approximately 2 percent sugar.

Often, yeast is added to help the sugar induce more effervescence in the second fermentation.

The Champagne region of France is a delimited zone approximately 90 miles east of Paris—it is about the size of Brooklyn. Champagne is the northernmost section of France still capable of producing wine grapes. The soil of Champagne is chalky, which gives the wine its distinctive flavor. The vines used for champagne are the same used in Burgundy, to the south of Champagne, but the taste is quite different.

You may have noticed that vineyard names, so prominently displayed on most wine bottle labels, cannot be found on champagne. The reason is simple: Champagne is still, centuries after Dom Perignon, a blend of different grapes, grown by different vineyards. The only three grapes found in French champagne are pinot noir and chardonnay (the same varieties grown in the Burgundy region) and pinot meunier, a black grape.

What country is the world's largest producer of sparkling wine? Not France, but Germany. German sparkling wine is usually called *sekt* or *schauwein* and certain sparkling Rhine wines are called *Sparkling Hock*. Spanish sparkling wine is never called champagne, but *spumanti*.

Clearly, the word *champagne* has such panache that any sparkling wine would love to be associated with it. California and New York wines are labeled as champagne even though they are by definition not products of the Champagne region.

The European wine producers are not reticent about labeling their wares champagne out of modesty or circumspection but because of long-standing treaties with France that prohibit all non-Champagne-produced sparkling wine from bearing the name champagne. Spain was the first country to formally consent to this agreement, in 1875, with the Treaty of Madrid, but all Common Market countries have agreed to abide by this provision. Of all the wine-producing countries of any consequence, only the United States, Australia, and Soviet Union refuse to take off the word *champagne* in their wine labeling.

Champagne labeling does not go unmonitored in the United

States, however. The Bureau of Alcohol, Tobacco and Firearms (BATF) classifies all wine. Labeling requirements on generic wines are rather loose, but on nongeneric wines, such as Medoc, or semigeneric wines, such as champagne, the rules are much more stringent. No wine can be labeled differently in this country than in its country of origin. And with domestic champagnes, the words *New York State*, for example, must appear in the same type size as *Champagne*.

Still, modest attempts at deception persist. Some domestic champagne makers have attempted shortcuts in the methods of producing the wine. For example, when a domestic champagne label says "fermented in the bottle," there is a good chance that the wine was not fermented in the bottle at all but in a large vat. When a label says "fermented in *this* bottle," you can be assured that more traditional fermentation methods have been used. Can there be any other reason for the "this/the" switch than an attempt to deceive the public?

Champagne has an image in almost every country as a beverage used for celebrations and special occasions. A good spumanti may be better than a bad champagne, but you would have a hard time convincing status-seeking customers. If the other European countries hadn't long ago ceded the right to the use of the term, we would probably be flooded with three-dollar champagnes on our supermarket and liquor store shelves.

Why do we tie shoes to the back of newlyweds' cars?

Most of us believe we tie shoes on the back of newlyweds' cars as a form of hazing, but our "practical joke" is actually an adaptation of an ancient ritual whose significance has been lost to its modern-day practitioners. Before the wide use of the automobile, members of the wedding party threw shoes at the departing bride and groom, but the meaning of the rituals is the same.

Although Egyptians may have started the practice, the answer to this Imponderable was first recorded as a custom of the Israelites in the Bible. Jews used a shoe as a symbol of ownership, signifying possession or authority over property or persons. Instead of attaching a seal or signature to a contract, a Jewish seller would remove a shoe before witnesses to signify the closing of a business deal and hand it to the buyer. Shoes were actually placed on tracts of land to symbolize ownership of the property.

The throwing of a shoe was also a gesture of renunciation.

In the Bible, God says, "Over Edom I cast out my shoe," meaning that Edom no longer received His protection.

In biblical days, women were a form of property, and not always a valuable form at that. The brother of a childless man had first-refusal rights over his widow; she was not allowed to marry again until her brother-in-law formally rejected her. This rejection was acted out in public. If the brother-in-law did not want her, the widow took off his shoe and spat before him. Once he released his shoe, he abandoned any claim to possession of her. According to the Bible, when Ruth's first husband's brother refused her, he delivered his shoe to Boaz as a renunciation of his claim over her. This ritual enabled Ruth to marry Boaz and retain her sterling reputation.

Christians adopted the shoe as a symbol of ownership. Historians have cited that when the Emperor Vladimir wanted to marry the daughter of Raguald, she refused with the gracious comment, "I will not take off my shoe to the son of a slave." Martin Luther was reputed to have placed the groom's shoe at the head of the bed to indicate who was boss to the bride. In Anglo-Saxon marriages, the father of the bride gave her shoe to the bridegroom, who then touched his wife-to-be on the head with it, to signify the passing of ownership from father to son-in-law.

Europeans eventually changed these shoe rituals and added new meanings. At first, parents threw shoes at the bride as a renunciation of their authority over her. Soon, other attendees at weddings threw shoes as well, and the practice became a good-luck ritual rather than a consummation of a business transaction.

In England, particularly, a lot of shoes were thrown all over the place; just about anyone embarking on a trip was worthy of getting pummeled for the good luck it supposedly bestowed.

Ben Jonson wrote:

Hurl after me a shoe,
I'll be merry whatever I'll do.

And Tennyson, in his "Lyrical Monologue," states:

For this thou shalt from all things seek
Marrow of mirth and laughter,
And wheresoe'er thou move, good luck,
Shall throw her old shoe after.

Sailors, always a superstitious lot, welcomed shoes tossed at them upon embarkation. Others burned old boots before leaving on a journey.

Presumably some time before the price of leather skyrocketed, brides used to throw off their right shoe and fling it amid the competing hordes of unmarried guests. As with the latter-day bouquet, the catcher of the shoe was slated to be the next to wed. Today, we seldom see shoes being thrown at newlyweds, but the custom of tying shoes to the car is clearly a stepchild of these antecedents.

The real Imponderable is: Why do these rituals endure when their original meaning has been obscured? Some historians speculate that many of the wedding rituals date back to the good old days when warriors stole their wives from under the noses of irate fathers and brothers and when the best man was really a bodyguard, protecting the groom's flank. The throwing of shoes, rice, and other objects can then be seen as a mock reenactment of the hostility between claimants on the bride. This explains why the bride and groom, who have been living it up at their postnuptial bash, all of a sudden dart toward their getaway car—it is a remnant of the time when bride and groom genuinely had to "make a run for it." Freudians argue that these wedding rituals are ways for the families involved in the wedding to sublimate disturbing Oedipal feelings.

Imponderables prefers a middle position. Weddings, like all rites of passage, are a way for a society to promote change and to mark transitions that might be traumatic without the aid of ritual. Even though most fathers, for example, no longer view their daughters as chattel, many still feel a deep sense of loss in giving away the bride. The whole wedding, to parents, serves as a renunciation ritual. In ancient times, fathers renounced their property rights; now parents, ambivalently, use the wedding to announce to society that they have renounced their authority over their daughter's life. The "tears of happiness" at weddings (see Imponderable "Why do we cry at happy endings?") are usually tears of sadness, grieving the loss of an intimate, mutually dependent relationship that will never return.

Why don't professional wine tasters get drunk on the job?

Because they don't swallow any wine. Watch oenophiles drink. They sniff the bouquet of the wine, circling their prey like a bloodhound. Then, not ones to gulp down anything, they slosh their wine in their mouths like a dental patient before rinsing (indeed, the slang term for such tasting is *gargling*). To the wine fancier, these rituals are necessary to savor the beverage; to the uninitiated, they are just damn pretentious.

Professional tasters don't give a whit about how their tasting rituals appear to others. Their job depends upon their ability to make fine distinctions between very subtle differences in taste, and they can perform their job perfectly well without swallowing a drop.

There are about ten thousand taste buds in the human mouth. Taste buds are receptors distributed mainly on the tongue, but also in the mouth and throat. Taste buds can discriminate only four different tastes: saltiness, sweetness, sourness, and bitterness. The taste buds that are receptive to these four qualities are located in specific areas. Sweetness and saltiness buds are located at the tip of the tongue; sourness at the sides; and bitterness on the back anterior of the tongue. One of the reasons tasters slosh wine around in their mouths is because the movement ensures that the wine will hit each of the types of taste buds that signal the brain to recognize all four tastes. Some theorists even believe that the location of the taste buds accounts for other drinking habits. We may sip wine, for example, because the slow intake allows us to better taste the sweetness of the wine lingering on the tip of the tongue, while we may shoot back shots of whiskey and beer because bitter taste buds are located primarily on the back of the tongue. In either case, the taste has registered on the brain before the drink has been swallowed.

As we age, our taste buds become less and less sensitive, and the buds at the tips of our tongues are the first to go. This explains why elderly people often regain the sweet tooth of their childhood—they need more sugar to taste the same sweetness that they liked as young or middle-aged adults.

The sniffs of the professional taster are as important as the gargling in detecting defects in a wine, but the sense of smell continues to play a crucial part in tasting after the drink is in the mouth. Odorous components of wine rise through the throat into the nose. The professional taster judges the total sensory quality of the wine, and it is next to impossible, even for the professional taster, to separate taste from odor when judging flavors in complex beverages such as wine.

Most of us judge food based on our olfactory sense. Perhaps some of you won science-fair competitions by having suckers try to tell the difference between apples and raw potatoes when their eyes were closed and nasal passages were blocked. Both are crunchy; both are sweet. Without the nose (and eyes) it is difficult to tell the difference between a cola and a lemon-lime soft drink. One of the reasons it is difficult to separate smell from taste is that the centers for these two senses are located right next to each other in the temporal lobe of the brain. Scientists haven't determined exactly how they interrelate, but their commingling is assumed.

Many sensations that we assume are related to taste are in fact "touch" sensations. The tanginess of a lemonade or a sharp cheese is neither a matter of taste nor smell. Hot dishes such as curries or mustards cause chemical reactions rather than taste sensations. We use our gums, lips, hands, and eyes to taste and evaluate the texture of food and drinks.

Our sense of taste is an extremely blunt one, possibly because nature intended taste to provide for the survival of the species rather than its aesthetic enhancement. Yet it is difficult not to admire professional tasters. Not only are they under pressure to detect characteristics in food and drink that the average person cannot, but they aren't given the soul-satisfying pleasure of swallowing their subjects.

Why is a mile 5,280 feet?

The word *mile* comes from the Latin word for one thousand, *mille*. So why aren't there 1000 feet in a mile? The Romans measured a mile as 1000 Roman paces (i.e., 2000 steps) by their marching soldiers; since the average marching stride was about five feet, the Roman mile was almost exactly 5000 feet.

The British were partial to the furlong, a unit of measurement that is used now primarily at horse racetracks. Even before written records of land were kept, British farmers built stone walls to demarcate fields whose length was standardized—the plowmen dug furrows the equivalent of 220 modern yards. *Furlong* was the slurred pronunciation of furrow-long, and the furlong became the designation for 220 years.

When Britain adopted the mile, farmers insisted that it be tied to their basic unit of measurement, the furlong. The Roman mile consisted of a little more than seven and one half furlongs. Rather than change their beloved furlong, the British changed the length of the Roman mile. Instead of 5000 feet, the British mile became eight furlongs, or 1760 yards, the exact measurement we now use today—5,280 feet. This also explains why the length of horse races is described in furlongs: Racetracks are exactly eight furlongs in circumference—a one-mile race is exactly one lap, and each furlong is an easy to visualize one-eighth of a mile.

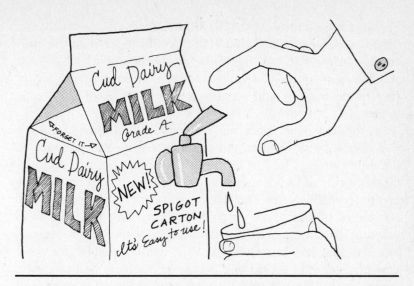

Why are milk packages so difficult to open and close?

In the good old days, milk was marketed in glass bottles and, more than likely, delivered to your home. When the family was finished with the bottles, the used glass containers were returned to the delivery person for recycling by the milk company.

In many ways, glass was the ideal packaging material for milk. Glass is inert, so it does not give off any taste of its own, and glass is durable enough to withstand the many recyclings it underwent.

When the economics of distribution led to the inevitable downfall of home delivery in most areas of the country, paper replaced glass in supermarket dairy cases. Round glass containers took up more space on the shelf than the rectangular paper milk cartons, and supermarkets were always interested in any packaging that helped preserve valuable shelf space. Paper was also considerably lighter than glass, making delivery easier.

Most importantly, paper was much cheaper as a raw material for the dairies than glass. Paper could be sent to the processing plants flat, and machines could assemble the cartons and fill them quickly. Paper containers eliminated the costly step of rinsing out and sterilizing previously used glass containers.

Untreated paper does not respond well to moisture, so it was necessary to insulate the paper to ensure the integrity of the milk carton. The first "solution" was wax, but that chipped off the inside of the container and ended up in the cereal bowls of unhappy children. The dairy industry soon discovered the miracle of plastic. Today's paper milk cartons are lined with polyethylene, a material that does not peel away from the carton and that has proven to be safe. Unfortunately, the very polyethylene that solved one packaging problem helped create another—the blasted milk cartons were difficult to open and equally as difficult to make airtight when closing.

There is simply no way to close the usual paper milk carton and be sure that air is not entering. This is one reason why, in many areas, plastic containers with screw-on caps have replaced paper containers for half gallons and gallons of milk—although some air may infiltrate a quart of milk, a quart is likely to be used so quickly that it is unlikely to spoil before being consumed.

Why doesn't the milk industry simply convert all of its containers to plastic, which are a breeze to open and close securely? The cost of the material is approximately the same for both paper and plastic, between 10 and 11 cents for a half-gallon container, according to Frank Finnegan of New York's Dellwood Foods. But there are practical advantages to paper containers. Assembly lines can operate much faster on paper than on plastic, saving time and, therefore, enough money for the industry to prefer paper packaging, all other things being equal.

Some people complain that there is an off taste created by oxidation in plastic containers, although this issue is hotly disputed by the plastics industry. Another controversial issue that works in favor of paper is whether or not plastic responds as well to heat and light as treated paper. Some professors at Cornell University reported that plastic containers can lose as much as 10–15 percent of the vitamin value of the milk they hold when exposed to heat or strong light.

If consumers showed a strong preference for plastic containers, Finnegan indicated that the industry would probably respond to their wishes, although the scattered complaints he now receives indicate that most milk buyers have become reconciled to cartons' less than perfect design.

Another reason the milk industry might be reluctant to give up its traditional paper packaging is that the humble gabled carton is one of the most instantaneously recognizable items on the supermarket shelf. Is there any inherent reason why yogurt containers should be tall and thin and cottage cheese containers short and squat? Not really, but we would not need labels or brand names to pick out which container held yogurt or cottage cheese, or catsup or mustard for that matter.

The tall, streamlined look of milk cartons, if forsaken for a more spherical plastic container, might not only be more difficult to stack in quantity on the supermarket shelf, but less distinctive looking. Any designer can create packaging that differentiates it from others in the same product category, but the risk in radically changing the packaging—providing consumers with false cues about what product they are going to buy—is simply not worth the gamble.

As it is, most of us could probably pick our favorite type of milk while sleepwalking. If we see a red and black container, it is likely to be homogenized milk; a blue and white package means low-fat or skim milk; and a green carton denotes buttermilk. If we see plastic containers, we can assume they are holding gallons.

The dairy industry hasn't deliberately tried to create packages impossible to open and close. It has simply hit upon packaging that is cheap, safe, and convenient from its point of view.

It is extremely difficult and extremely expensive to make paper containers that can be opened easily and sealed tightly. Unless the industry feels that its packaging affects its bottom line negatively (in which case it would switch to plastic rather than "new, improved" paper), it has no reason to change.

How does Kraft get "five ounces of milk in every slice" of American Singles?

"Five ounces of milk in every slice" is an effective advertising slogan for at least three reasons. First, it reinforces the desired wholesome image for Kraft Singles, especially since the product is largely consumed by children and usually purchased by their nutritionally concerned parents. Second, it identifies a value inherent in the product. A slice of Kraft Singles costs approximately 15 cents; if it contains 5 ounces of milk, it's not such a bad deal. Third, it provides what advertisers call a "unique selling proposition." Simply stated, if a particular brand can make a distinctive claim for its product, it will carve out a niche that cannot be broken by the opposition as long as the unique selling proposition is broadcast to enough people for a long enough time. Kraft has no problems in this regard, as it dominates the processed cheese-food category and overwhelms any other cheese maker in advertising reach.

There is one corollary to the unique selling proposition principle that doesn't get discussed too often: If you can make a claim for your product that is true for other brands as well, you can still "own" the claim—as long as you appropriate the promise in advertising first and proclaim it loudly enough. Ivory Soap's claim, for example, to be $99\frac{44}{100}$ percent pure may be matched by other soaps, but no other manufacturer would be stupid enough to base an advertising campaign on the notion that it was $99\frac{63}{100}$ percent pure. Simply by bringing up the issue, consumers would think of Ivory instinctively, not even hearing the rival product's claim.

Other cleaners besides Ajax are "stronger than dirt," but no rival would want to trumpet the claim. Ajax has already staked out the territory.

Kraft's "five ounces" campaign may be a unique selling proposition as an advertising slogan, but there is nothing mystical about how Kraft reduces a whole bunch of milk into one little slice—it is, actually, the story of how all cheese is made. The genius behind the Kraft campaign, however, is that the guarantee of the quantity of milk in each slice reassures skeptical consumers about the wholesomeness of a "process cheese food."

Kraft was kind enough to detail the process of how it produces a slice of Singles, and the description by consumer representative Jane C. Venters was so clear, we'll quote it verbatim:

> In all cheesemaking, ten pounds of milk are required to make one pound of natural cheese. Milk by its very nature contains approximately 87 percent water. During cheesemaking the water and other substances are separated out of the milk and removed. The portion removed is called the whey, and it accounts for about nine pounds of the original ten pounds of milk. The remaining solid mass, which has been formed by coagulating the milk, is called the curd. The curd is pressed and formed into what we know as natural cheese, and represents about one pound of the original ten pounds of milk.
>
> Kraft American Singles Pasteurized Process Cheese

Food is made from natural cheese. As we said above, ten pounds of milk are required to make one pound of natural cheese. Therefore, with this information, you can calculate that 5-oz. of milk will yield ½-oz. of cheese. A ¾-oz. slice of American Singles contains about ½-oz. of natural cheese. The remaining ¼-oz. contains other ingredients as listed in the ingredient statement on our packages.

As you can see from the above information, there are indeed 5-oz. of milk in a slice of Kraft American Singles Pasteurized Process Cheese Food. We assure you that our research scientists verified the accuracy of this statement before the commercial was produced and aired. All advertising campaigns are carefully documented, and are thoroughly reviewed to insure their truthfulness and validity.

What are the remaining ¼-oz. of ingredients that comprise a slice of Kraft Singles? Here are the contents of the label: "Natural Cheddar Cheese, Water, Whey, Sodium Citrate, Whey Protein Concentrate, Skim Milk, Milkfat, Sodium Phosphate, Salt, Sorbic Acid as a preservative, Annatto and Apocarotenal (color)." Much of the other ¼-oz. of each slice, then, is clearly made up of water, both in the form of straight H_2O and whey byproducts, which are the watery stuff normally taken out of natural cheese. These additional ingredients are why Kraft cannot call its Singles cheese, but is forced to resort to tongue twisters like Pasteurized Process Cheese Food. The meaning of each of these four words is specified by the Food and Drug Administration.

How do the police make crowd estimates?

If you engage in a war, folks at the home front want to know the body counts. Throw a parade or a riot, and people want crowd estimates. It's human nature to want to judge the failure or success of an enterprise by quantifying it.

The unenviable task of making crowd estimates usually falls on the local police department, and parades are usually the occasions for these estimates. The most famous parade in the United States, the Tournament of Roses parade, held in Pasadena, California, every New Year's Day, has, since 1930, consistently estimated its attendance at from 1 to 1.5 million. The Pasadena police would be quite happy not to make crowd estimates, but the press needs figures (it just doesn't sound right to start a newspaper story about the parade by saying, "A whole bunch of people showed up in Pasadena . . ."), and politicians need to measure the success of the parade in order to boast of their accomplishment.

But how are these estimates made? Imagine the logistical nightmare of trying to count heads at a ticker tape parade in lower Manhattan, with its asymmetrical streets, floating debris (obscuring vision), and the staggering numbers involved.

For several years, Michael Guerin, the special events public information officer for the City of Pasadena, has had the responsibility of figuring out the attendance at the Tournament of Roses parade. Guerin flies over the parade site in a helicopter. Obviously, he doesn't count heads. From his years of experience, he knows what 104,000 people look like bunched up together, for that is the capacity of the Rose Bowl, home of the football institution that follows the parade. Using the Rose Bowl crowd as a benchmark, Guerin tries to conceptualize the 100,000 + people he has seen in the circular stadium into the linear crowd along the parade route. This can't be a precise measurement; after all, the parade route spans exactly five and a half miles, and he must also count spectators who look at the floats in the formation areas, where they are assembled, as well as the post-parade area where the floats are put on display.

Since the population of Pasadena is well under 200,000, local officials are used to skepticism about their estimates of 1,000,000 plus spectators, and were challenged in 1983 by Peter Apanel, founder of the Doo-Dah parade, a spoof of the Tournament of Roses parade. Dubious about official estimates, Apanel commissioned photographers to shoot 442 sequenced snapshots of spectators lining the Doo-Dah parade (which that year, police and Apanel agreed, attracted more than 50,000 viewers) at fixed intervals. Apanel then counted every single person in those pictures and extrapolated density levels applicable to the New Year's parade. Although he claimed that for two-thirds of the route, the shorter Doo-Dah parade had as much or more spectator density, he multiplied the density level by two when estimating the Tournament of Roses parade crowd. By factoring in the fans sitting in the reserved bleachers and the longer route of the Tournament of Roses parade, Apanel insisted that the police estimate was way off—that no more than 360,000 people could have attended the parade in 1983, or about one-fourth of the police estimate.

Robert Gillette, a reporter for the *Los Angeles Times*, did a little figuring of his own. He measured the depth of the standing room area at 23 feet, marked on one side by the blue line behind which all spectators must stand and on the other side by the buildings at the back of the crowd. Multiplying this standing room area by the 5.5-mile parade route, Gillette calculated that the parade route provided 1,336,000 square feet (not all of this space was occupied, since attendees toward the back can and often do move about freely, but Gillette did not factor in unused space). He then assumed that each attendee occupied two square feet (one foot thick and two feet wide). Dividing the 2 square feet into the 1,336,000 square feet, Gillette arrived at the figure of 668,000 as the maximum number of people that the Tournament of Roses parade route could accommodate.

Other skeptics have arrived at different estimates, including some Pasadena-based California Institute of Technology professors, who put a half million as the maximum number. Guerin, however, feels confident in his approximation and notes that in the helicopter, he is always amazed how fluid the pedestrian traffic is. In the early morning, it always looks like attendance is bad, but somehow new people keep appearing. Guerin added that although the official Pasadena estimate is hardly precise, it's as good a guess as anyone else's and that police don't receive any kind of special training or education in crowd estimation.

Imponderables spoke to New York officials about how they make crowd estimates. In most cases, the task is left to the local police precinct where the parade takes place.

The police make stabs at accuracy, but it is no more a science in New York than in Pasadena. The most popular technique for Fifth Avenue parades (such as the St. Patrick's Day parade or Columbus Day parade) is to count the number of rows of spectators behind the blue wooden barriers that are placed on each side of the street. Each barrier is fourteen feet long. Assuming that the population behind each barrier will reflect the parade route as a whole, the police estimate how many spectators fit into the square footage available—in essence, they duplicate the methodology of Doo-Dah founder Apanel without using photographs, and simply assume that density levels will not vary greatly at different points in the parade route.

Another, more ingenious method of estimating crowd size is by examining the quantity of artifacts they leave behind. To say it less delicately, one way of counting a crowd is to weigh how much garbage it leaves behind. Since sanitation trucks are weighed electronically at the disposal site, it has always been an easy matter to measure the amount of debris left after New York's famed ticker tape parades down the "canyon of heroes."

Counting garbage is not a perfect scheme for measuring crowd sizes, however. The Macy's Thanksgiving Day Parade, for example, is notorious for its large attendance but pitiful lack of garbage, which rarely surpasses ten tons. Even the Tournament of Roses parade weighs in at around a measly forty tons a year. These parades are pikers compared to the ticker tapes, but the latter have the advantage of artificial inflation—in recent years, some parade committees have actually imported shredded paper from out of the city to be thrown at passing heroes. Not much ticker tape is thrown anymore. Computer printouts are the replacement. More and more skyscrapers are "climate controlled," with windows incapable of being opened, reducing the opportunities for many to contribute to the mess. All of these factors make it difficult to correlate crowd size with quantity of garbage, but the New York City Department of Sanitation is besieged with requests for the garbage count, and most observers feel there is some connection between the amount of paper thrown and the frenzy and enthusiasm of the celebrants.

For Casey Kasem fans everywhere, here are the top five garbage parades of all time in New York:

5. 1969 Mets parade—1254 tons.

4. Iranian hostages' return, January 30, 1981—1262 tons.

3. Douglas MacArthur's return, April 20, 1951—3249 tons.

2. John Glenn, March 1, 1962—3474 tons.

1. V-J Day, August 14, 1945—5438 tons.

By all acounts, the V-J Day parade was the most spirited and most heavily attended.

Every single source I spoke to about this Imponderable conceded that precision in estimating crowds was impossible and the task itself of less than earth-shattering importance. None was trained to execute this task. And all of them felt that newspapers and politicians would force them to continue with the madness.

Why don't we ever see baby pigeons?

Pigeons, or rock doves, as your high school biology teacher would more properly call them, are known in the birding trade as ledge nesters. In the wild, they build nests on cliffs, canyons, and rocky terrains. But pigeons are just as comfortable using man-made structures such as bridges and ledges of buildings. You won't find pigeon nests in trees.

Baby pigeons have an extremely high metabolism, eating a large proportion of their body weight every day. They grow so fast that by the time their mothers kick them out of the house (usually within one month of birth), baby pigeons, like all birds, are close to or have already achieved full size. When you think you see babies loitering with their parents, you are probably spotting two different species of birds.

The only way to distinguish between an immature and mature pigeon is to examine its plumage. Younger birds tend to have raggedy feathers, especially at the ends of their tail feathers. Although pigeons have varying colorations, mature birds tend to have brighter feathers.

Why do they sell 40-, 60-, 75-, and 100-watt bulbs?

Imponderables could never understand why manufacturers made bulbs with these wattages. Wouldn't it be more aesthetic, more symmetrical, to make 50-, 75-, and 100-watt bulbs? Was there some technological or marketing reason for the present configurations?

After doing quite a bit of our own research and enlisting the help of several lighting companies, no one could provide a definitive explanation for the initial selection of the enduringly popular 40- and 60- and 100-watt bulbs. But James H. Jensen, at General Electric, supplied a fascinating history of the first light bulbs manufactured for home use.

Thomas Edison's first lamps were sized to equal the light output of the gas mantles that were then in use. In fact, when electric lamps were first manufactured, bulbs were not rated in watts but in candlepower, a measurement of their light output. Jensen theorizes that the 40-, 60-, and 100-watt bulbs were designed to correspond to the light output (or candlepower) of lanterns already in use, for they were the first three wattages marketed when the tungsten filament was introduced in 1907. These were the only three sizes offered until 1916, when 50- and 75-watt bulbs were also sold.

Jensen notes that a major reason for the introduction of the 50- and 75-watt sizes was that these bulbs were literally given away free to customers by electric utility companies, such as Commonwealth Edison and Detroit Edison. It might be difficult to conceive of the need to sell the public on using a source of energy, but it was not always a foregone conclusion that the country would quickly and universally adopt electric energy in its homes. Electricity was waging a marketing fight with gas, just as the natural gas and home heating oil companies today are trying to shake off excess supply by urging consumers to switch energy allegiance.

It does not take a cynic to figure out that the reason the electric companies supplied 50- and 75-watt sizes was to replace the 40- and 60-watt size with more hearty electricity eaters. Nor does it take an advanced mathematician to calculate that both new sizes were exactly 20 percent bigger than their predecessors, another indication that the new sizes were more likely to be the brainchild of an accountant rather than a response to customer preference.

Eventually, the electricity industry realized that it had won the war and that consumers had no intention of reverting to gas lanterns. Free light bulbs were taken away for the same reason that trading stamps were yanked from service stations during the gas crisis—why give away something for nothing when it doesn't increase usage of your product?

Without benefit of subsidization by the electric utilities industry, former beneficiaries of the free bulbs went back to the retail counter and bought the bulbs that the rest of the country had always shown a preference for, the 40- and 60-watt bulbs.

Although a few manufacturers still make the 50-watt bulb, it has just about disappeared. General Electric, for example, has dropped it, but retains the 75-watt bulb, which sells nowhere near as much as the old reliables.

Why do women open their mouths when applying mascara?

More people were contacted to help answer this Imponderable than any other. We asked cosmeticians; we asked cosmetics companies; we asked plastic surgeons; we asked women off the street; and we asked women on the street. We had some very intelligent people standing in front of mirrors opening their mouths and pretending to apply mascara, looking to see if their eye muscles moved. Everybody seemed to have the answer. It just wasn't the same answer.

There were staunch supporters of six different theories:

The "Edith-Ann" Theory

Remember Lily Tomlin's desperately serious little girl in the huge rocking chair? Whenever confronted with a difficult task, Edith-Ann would grit her face, open her mouth, and stick out her tongue. Many children follow the same pattern when trying to concentrate. The Edith-Ann theory proposes that women open their mouths when applying mascara for the same purpose—to better concentrate on the delicate balancing act of working around the eye. Of course this theory doesn't explain why mouths remain closed during other delicate operations—like swabbing ears, putting on hair rollers, or painting toenails.

The "I Didn't Want to Do It" Theory

Proponents of this theory claim that opening the mouth is merely a reflex action, totally involuntary. This theory would explain the phenomenon but has the flaw not only of having no evidence to support it, but being proven untrue by medical research.

The "No More Blinking" Theory

Blinking is the enemy of mascara application, so any trick to avoid blinking would come in handy. Many women supported this theory but had a difficult time sustaining the argument when, upon demand, they were unable to stop blinking when their mouths were open.

The "Stupid Imponderable" Theory

Although no one denied the fact that most women open their mouths when applying mascara, some people held the opinion that the act performed no useful function and that trying to research this Imponderable was an incredible waste of time. There is much merit to this theory, but it is obviously incorrect, since if it were correct, this Imponderable would not appear in the book.

The "Mellow-Out" Theory

The largest number of sources subscribed to this theory. Opening the mouth is a relaxation response, they argued. Unlike larger muscles in the arms and legs, facial muscles are not very autonomous—they tend to work in groups. Opening the mouth seems to relax the whole face, making it easier for the woman to endure the laborious process of mascara application. The "Mellow-Out" theory totally contradicts . . .

The "Tighten Up" Theory

This theory postulates that opening the mouth tightens muscles and puts the skin on the eyelids on a stretch, making it easier to apply mascara for much the same reason that men make jaw contortions (often including opening the mouth) to make shaving the neck easier. Using the same principle as their arch-rival Mellow-Out adherents, Tighten Ups believe that opening the mouth tightens all the facial muscles. By stretching the skin slightly, there is an increase of exposure on the eyelid, making application of mascara that much easier.

Imponderables comes down on the side of the Tighten Up theory, not only because we are fond of the song bearing the same name, but because we found two eminent plastic surgeons whose livelihood depends upon knowing the effect of opening the mouth on the eye. Both Dr. Gerald Imber, of New York City, and Dr. Tom Flashman, of Beverly Hills, California, told us the same story. When they are contemplating blepharoplasty (eyelid surgery), particularly surgery on the lower lid, the doctors have their patients open their mouths. This tenses the lower lid and even pulls the lids a little more apart from each other. Opening the mouth and stretching the skin on the eyelid often reveals that more or less skin can be removed than originally contemplated. As the decision about precisely how much skin to cut off during blepharoplasty can make the difference between a cosmetically successful or lousy "lid job," having their patients perform this ritual is a minor but essential diagnostic tool.

In researching this Imponderable, we perused many books on makeup. Several bestsellers, such as *Mary Kay Guide to Beauty* and Carol Jackson's *Color Me Beautiful*, included advice to open the mouth when applying mascara, but not one said why. Not one of the cosmeticians we spoke to at department stores knew the correct answer to this question, yet they all recommended the procedure.

Women know how to apply mascara properly even if they can't explain why their method works. Is it instinct? Is there an underground educational network? Fodder for another Imponderable, perhaps.

Index

Master Index of Imponderability

Following is a complete index of all ten Imponderables® books and *Who Put the Butter in Butterfly?* The bold number before the colon indicates the book title (see the Title Key below); the numbers that follow the colon are the page numbers. Simple as that.

Title Key

 1 = Why Don't Cats Like to Swim? (*formerly published as* Imponderables)
 2 = Why Do Clocks Run Clockwise?
 3 = When Do Fish Sleep?
 4 = Why Do Dogs Have Wet Noses?
 5 = Do Penguins Have Knees?
 6 = Are Lobsters Ambidextrous? (*formerly published as* When Did Wild Poodles Roam the Earth?)
 7 = How Does Aspirin Find a Headache?
 8 = What Are Hyenas Laughing at, Anyway?
 9 = How Do Astronauts Scratch an Itch?
10 = Do Elephants Jump?
11 = Who Put the Butter in Butterfly?

Art pencils, grades of, **7**:73

Aspirin
  headaches and, **7**:100–102
  safety cap on 100-count bottles of, **4**:62

Astrology, different dates for signs in, **4**:27–28

Astronauts and itching, **9**:208–216

"At loggerheads," **11**:104–105

Athletics, Oakland, and elephant insignia, **6**:14–15

"Atlas," **11**:154–155

ATMs
  swiping of credit cards in, **10**:138–141
  swiping versus dipping of credit cards in, **10**:141–142
  transaction costs of, **3**:102–103

"Attorney-at-law," **11**:103–104

Auctioneers, chanting of, **9**:201–204

Audiocassette tape on side of road, **7**:250–251

Audiocassette tapes on roadsides, **9**:300

Audiotape, versus videotape technology, **3**:136–137

Audiotape recorders
  counter numbers on, **4**:148–149
  play and record switches on, **5**:23–24

Automobiles
  batteries and concrete floors, **10**:234–236
  bright/dimmer switch position, **7**:44–46; **8**:258

bunching of, on highways, **7**:247

cardboard on grills, **6**:188–189; **7**:245–246

clicking sound of turn signals, **6**:203

cockroaches in, **7**:3–4; **8**:256–257; **9**:298

Corvette, 1983, **9**:137–139

cruise controls, **8**:124–125

day/night switch on rear-view mirrors, **4**:185–186

dimples on headlamps, **8**:56

elimination of side vents, **6**:13–14

gas gauges in, **3**:5–6; **6**:273

headlamp shutoff, **7**:92–93

headlights and deer, **6**:212–214

holes in ceiling of, **5**:179

key release button on, **5**:169

keys, ignition and door, **3**:141–142

"new car smell," **5**:63

oil loss after oil change, **7**:240–241

oil, grades of, **3**:182–183

rear windows of, **5**:143–144

rentals, cost in Florida of, **4**:24–25

side-view mirrors, **2**:38–39

speed limits, **2**:143

speedometers, **2**:144–145

tire tread, **2**:72–74

weight of batteries, **5**:101–102

white wall tires, **2**:149

windshield wipers, versus buses, **7**:28

Autopsies of executed criminals, **8**:13–15

"Ax to grind," **11**:70

Babies
blinking, **6**:158–159
burping, **10**:123–124
hair color, **10**:209
high temperatures, tolerance of, **4**:103–104
sleep, **6**:56–57

Baby corn, in supermarkets, **10**:186–187

Baby pigeons, elusiveness of, **1**:254; **10**:251–253

Baby Ruth, origins of name, **8**:84; **9**:288–289; **10**:264–265

Baby shrimp, peeling and cleaning of, **5**:127

"Back and fill," **11**:2–3

Back tapping during physical exams, **3**:145–146

Backlogs in repair shops, **4**:45–47

Bad breath, upon awakening, in the morning, **4**:52

Badges, marshals' and sheriffs', **5**:73–74

Bagels, holes in, **3**:90–91; **8**:261

Bags under eyes, **3**:151

Bags, paper, jagged edges and, **6**:117–118

Baked goods
Pennsylvania and, **2**:121–122
seven-layer cakes, **6**:80–81
unctuous taste until cooled, **6**:151–152

Baked potatoes in steak houses, **6**:127–129

Ball throwing, sex differences and, **3**:42–44

Ballet, "on pointe" in, **8**:69–72

Balls, orange, on power lines, **4**:18–19

Balls, tennis, fuzz and, **3**:35–36

Balsa wood, classification as hardwood, **5**:85–86

Banana peels as slipping agents, **3**:198; **5**:228–229; **6**:250–252

Bananas, growth of, **2**:81–82

Band-Aid packages, red tear strings on, **1**:147–149; **6**:266

Bands, tardiness of, in nightclubs, **8**:184; **9**:248–254

Bands, marching, formation of, **8**:107–108

Bands, paper, around Christmas card envelopes, **6**:203–204

Banking
ATM charges, **3**:102–103
hours, **3**:100–101

Barbecue grills, shape of, **10**:99–102

"Barbecue," **11**:173

Barbie, hair of, versus Ken's, **7**:4–5; **8**:259–260

Barefoot kickers in football, **4**:190–191

Bark, tree, color of, **6**:78–79

Barns, red color of, **3**:189–191

Bars
mirrors in, **10**:14–17
sawdust on floor of, **10**:118–120
television sound and, **10**:12–14

Baseball
  black stripes on bats, **8**:104–106

  Candlestick Park, origins of, **9**:48–51

  cap buttons, **9**:171–172

  caps, green undersides of, **9**:172–173

  circle next to batter's box in, **3**:28

  dugout heights, **5**:14

  first basemen, ball custody of, **1**:43–44

  home plate sweeping, by umpires, **8**:27–31

  home plate, shape of, **5**:131

  Japanese uniforms, **10**:207–208

  "K," as symbol for strikeout, **5**:52–53

  pitcher's mound, location of, **5**:181; **9**:195–198

Baseball cards
  wax on wrappers, **5**:123

  white stuff on gum, **5**:122

Basements, lack of, in southern houses, **4**:98

Basketball
  24-second clock in NBA, **1**:29–31

  duration of periods in, **9**:65–69

Basketballs, fake seams on, **4**:155–156

Baskin-Robbins, cost of cones versus cups at, **1**:133–135

"Batfowling," **11**:1–2

Bathrooms
  group visits by females to, **7**:183–192; **8**:237–238; **9**:277–278

  ice in urinals of, **10**:232–234

  in supermarkets, **6**:157

Bathtub drains, location of, **3**:159–160

Bathubs, overflow mechanisms on, **2**:214–215

Bats, baseball, stripes on, **8**:104–106

Batteries
  automobile, weight of, **5**:101–102

  concrete floors and, **10**:232–234

  drainage of, in cassette players, **10**:259–260

  nine-volt, shape of, **6**:104; **7**:242–243

  sizes of, **3**:116

  volume and power loss in, **2**:76–77

"Battle royal," **11**:67

Bazooka Joe, eye patch of, **5**:121

Beacons, police car, colors on, **7**:135–137

"Bead," "Draw a," origins of term, **10**:168

Beaks versus bills, birds and, **10**:3–4

Beanbag packs in electronics boxes, **6**:201

Beans, green, "French" style, **10**:125–126

Beards on turkeys, **3**:99

"Bears [stock market]," **11**:106–107

"Beating around the bush," **11**:1–2

Beavers, dam building and, **10**:42–46

Chewing gum
lasting flavor, **5**:195–196
water consumption and hardening of, **10**:236–237
wrapping of, **8**:111–112
Chewing motion in elderly people, **7**:79–80
Chianti and straw-covered bottles, **8**:33–35
Chicken
cooking time of, **1**:119–121
versus egg, **4**:128
white meat versus dark meat, **3**:53–54
"Chicken tetrazzini," **11**:153
Children, starving, and bloated stomachs, **7**:149–150
Children's reaction to gifts, **8**:184; **9**:234–237
Chime signals on airlines, **7**:6–8
Chirping of crickets, at night, **10**:54–57
Chocolate
Easter bunnies, **2**:116
shapes of, **2**:24–25
white versus brown, **5**:134–135
wrapping of boxed, **8**:122–123
Chocolate milk, consistency of, **3**:122–123
"Chops," **11**:47
Chopsticks, origins of, **4**:12–13
"Chowderhead," **11**:72
Christmas card envelopes, bands around, **6**:203–204
Christmas tree lights
burnout of, **6**:65–66
lack of purple bulbs in, **6**:185–186; **9**:293; **10**:280

Cigar bands, function of, **4**:54–55
Cigarette butts, burning of, **5**:45
Cigarettes
grading, **6**:112
odor of first puff, **2**:238; **3**:223–226
spots on filters, **6**:112–113
Cigars, new fathers and, **3**:21–22
Cities, higher temperatures in, compared to outlying areas, **1**:168–169
Civil War, commemoration of, **3**:168–169
"Claptrap," **11**:73
Clasps, migration of necklace and bracelet, **7**:180; **8**:197–201; **9**:279–281
Cleansers, "industrial use" versus "household," **5**:64–65
Clef, treble, dots on, **10**:210–213
Clicking noise of turn signals, in automobiles, **6**:203
Climate, West Coast versus East Coast, **4**:174–175
Clocks
clockwise movement of, **2**:150
grandfather, **4**:178
number 4 on, **2**:151–152
Roman numerals, **5**:237–238
school, backward clicking of minute hands in, **1**:178–179
versus watches, distinctions between, **4**:77–78
Clockwise, draining, south of the Equator, **4**:124–125

Dictionaries
  pronunciation and, **10**:169–179
  thumb notches in, **5**:167–168
Diet soft drinks
  calorie constituents in, **6**:94–95
  phenylalanine as ingredient in, **6**:96
Dimples
  auto headlamps, **8**:56
  facial, **3**:23
  golf balls, **3**:45–46
Dinner knives, rounded edges of, **1**:231–232
Dinner plates
  repositioning of, **8**:184
  round shape of, **8**:162–164
Dirt, refilling of, in holes, **7**:48–49
Disc jockeys and lack of song identification, **7**:51–57
"Discussing Uganda," origins of term, **5**:246–247; **11**:145
Dishwashers, two compartments of, **6**:109–110
Disney cartoon characters
  Donald Duck, **3**:150
  Goofy, **3**:64–65; **7**:102
  Mickey Mouse, **3**:32
Disposable lighters, fluid chambers of, **6**:92–93
Distilleries, liquor, during Prohibition, **9**:54–56
Ditto masters, color of, **6**:133–134
"Dixie," **11**:147–148
Dixieland music at political rallies, **5**:203; **6**:234–236
"Dixieland," **11**:147–148

DNA, identical twins and, **10**:18–20
Doctors, back tapping of, **3**:145–146
Doctors and bad penmanship, **5**:201; **6**:221–225; **7**:232–233; **8**:235
"Dog days," **11**:17–18
Dogs
  barking, laryngitis in, **2**:53–54
  black lips of, **3**:38–39
  body odor of, **2**:40
  cavities and, **10**:277–278
  circling before lying down, **2**:2–3; **5**:238–239
  crooked back legs of, **8**:126–128
  Dalmatians and firefighting, **6**:270–271
  drooling in, **6**:34–35
  eating cat feces, **6**:35–37
  eating posture of, **6**:63–64
  head tilting of, **4**:198; **5**:215–217; **6**:258–259; **8**:237
  lifting legs to urinate in, **4**:35–36
  miniature, **6**:154–155
  poodles, wild, **6**:207–209
  rear-leg wiggling, when scratched, **6**:52–53
  "sic" command, **5**:51
  sticking head out of car windows in, **4**:60–61
  wet noses, **4**:70–73
Dollar sign, origins of, **7**:103–104
Dolls, hair of, **7**:4–5
Donald Duck, brother of, **3**:150

Donkey Kong, origins of, **10**:38–40

DONT WALK signs, lack of apostrophes in, **6**:75–76

Doors
double, in stores, **6**:177–180
knobs versus handles on, **7**:148–149
opening orientation of, in buildings, **4**:167
shopping mall entrances, **6**:180–181
THIS DOOR TO REMAIN UNLOCKED DURING BUSINESS HOURS signs, in stores, **6**:202

"Doozy," **11**:37

"Do-re-mi," **11**:29–30

Dots on cue balls, in pool, **10**:237–240

Double doors in stores, **6**:177–180

"Doubleheader," **11**:133

Double-jointedness, **10**:229–231

Double-yolk eggs, **3**:188–189

Doughnuts
tissues and handling in stores, **2**:164; **5**:240
origins of holes in, **2**:62–64

Downhill ski poles, shape of, **3**:69

Dr Pepper
origins of name, **5**:129–130; **6**:272
punctuation of, **8**:253–254

Drains, location of bathtub, **3**:159–160

"Draw a Bead," origins of term, **10**:168; **11**:134

Dreams, nap versus nighttime, **3**:124

Drinking glasses, "sweating" of, **9**:124–125; **10**:261

Dripless candles, whereabouts of wax in, **4**:182–183

"Driveway," **11**:64–65

Driveways, driving on, versus parkways, **4**:123

Driving, left- versus right-hand side, **2**:238; **3**:209–212; **6**:248–249; **7**:223; **8**:230–231

Drooling, dogs and, **6**:34–35

Drowsiness after meals, **6**:138–139

Drugstores, high platforms in, **8**:5–7

"Dry," as terms for wines, **5**:141–142

Dry-cleaning
French, **3**:164–165
garment labels and, **2**:59–60
One Hour Martinizing, **3**:28–29
raincoats and, **2**:216–217

Dryers, hand
in bathrooms, **10**:266–267
"off" switches, **8**:174–176

Ducks
lakes and ponds and, **10**:256
on Cadillacs, **5**:174–176

Duels, timing of, **5**:69

Dugouts, height of, **5**:14

"Dukes," **11**:137

"Dumb [mute]," **11**:131–132

"Dumbbells," **11**:131–132

Dust, ceiling fans and, **10**:263–264

E

as school grade, **3**:198; **4**:206–209

on eye charts, **3**:9–10

"Eagle [golf score]," **11**:139–140

Earlobes, function of, **5**:87–88

"Earmark," **11**:46

Earrings, pirates and, **9**:43–45; **10**:272–273

Ears

hairy, in old men, **2**:239; **3**:231–233; **5**:227–228

popping in airplanes, **2**:130–132

ringing, causes of, **2**:115–116

Earthworms as fish food, **3**:110–112

Easter

chocolate bunnies and, **2**:116

dates of, **4**:55–56

ham consumption at, **1**:151–152

"Easy as pie," **11**:172

Eating, effect of sleep on, **6**:138–139

"Eavesdropper," **11**:109–110

Ebert, Roger, versus Gene Siskel, billing of, **1**:137–139

Egg, versus chicken, **4**:128

Egg whites and copper bowls, **7**:99

Eggs

color of, **2**:189–190

double-yolk, **3**:188–189

hard-boiled, discoloration of, **3**:34

meaning of grading of, **4**:136–137

sizes of, **2**:186–188

"Eggs Benedict," **11**:154

"Eighty-six," origins of term, **10**:265–266; **11**:101–102

Elbow macaroni, shape of, **4**:28

Elderly men

pants height and, **2**:171–172

shortness of, **2**:239; **3**:229–231; **6**:250

Elections, U.S.

timing of, **6**:41; **8**:260–261; **9**:291–292

Tuesdays and, **1**:52–54; **3**:239

Electric can openers, sharpness of blades on, **6**:176–177

Electric drip versus electric perk, in coffee, **4**:35

Electric perk versus electric drip, in coffee, **4**:35

Electric plug prongs

holes at end of, **5**:94–95

three prongs versus two prongs, **5**:191

Electricity, AC versus DC, **2**:21–22

Electricity, static, variability in amounts of, **4**:105–106

Elephants

disposal of remains of, **6**:196–197

jumping ability of, **10**:27–29

Oakland A's uniforms, **6**:14–15

rocking in zoos, **8**:26–27; **10**:279

Elevator doors

changing directions, **8**:169–170

holes, **8**:170–171

Elevators
  overloading of, **5**:239
  passenger capacity in, **1**:23–24
"Eleventh hour," **11**:99–100
Emergency Broadcasting System, length of test, **1**:117–118
"Emergency feed" on paper towel dispensers, **8**:149–150
Emmy awards, origins of name, **2**:52
English, driving habits of, **2**:238; **3**:209–212; **6**:248–249; **7**:223; **8**:230–231
English horn, naming of, **5**:38–39
English muffins
  in England, **6**:200–201
  white particles on, **6**:49
Envelopes
  Christmas cards, paper bands on, **6**:203–204
  colored stickers on, **4**:83–84
  computer scrawls on, **5**:104–106
  red letters on back of, **5**:50
  return windows on, **2**:111; **5**:239–240
  size of, in bills, **6**:81–83
Escalator knobs, purpose of, **2**:42
Escalators
  green lights on, **3**:172
  rail speed of, **3**:171
Evaporated milk
  refrigeration of, **2**:114
  soldered cans and, **2**:114

Evolution, loss of body hair and, **2**:6–8
Exclamation marks in telegrams, **3**:76–77
Executions
  autopsies after, **8**:13–15
  hours of, **2**:34–36
EXEMPT signs at railroad crossings, **5**:118–119
Expiration dates
  on bottled waters, **9**:77–78
  on fire extinguishers, **4**:68
  on toothpaste, **4**:169–170
Exposures, number of, in film, **1**:153–154
Exterminators, hard hats and, **2**:51
Extremities, wrinkles on, **2**:112
Eye chart, E on, **3**:9–10
Eye closure during kissing, **8**:186–191
Eyeglasses and nerdiness, **7**:180; **8**:210–213
Eyes
  bags under, **3**:151
  floaters and, **3**:37
  pain, when tired, **6**:159–160
  position, during sleep, **6**:146
  rubbing of, when tired, **10**:103–105
  sneezing and, **3**:84–85

Fading, causes of, **9**:190–194
Fahrenheit scale
  increments in, **2**:55–56
  zero degrees, meaning of, **2**:55
Fainting and women, lessening of, **10**:219–222

FALLING ROCK signs, purpose of, **6**:72–74

Falsetto voices
   techniques for men producing, **4**:147
   women's inability to produce, **4**:147–148

"Fan [loyal partisan]," **11**:136

Fan speeds, order of, **10**:180–181

Farm plots, circular shape of, **7**:118–119

Fast-food restaurants, dessert and, **1**:218–221

Fat people and alleged jolliness, **4**:199; **5**:217–218; **6**:259; **7**:228

Fathers, new, passing out of cigars and, **3**:21–22

Fats, on nutrition labels, **6**:142–143

Faucets, water, **4**:191–192

"FD&C," meaning of, on food and shampoo labels, **4**:163

Federal Express, hours of employees of, **2**:109–110

Feet, swelling in airplanes of, **2**:31–32

Fences, height restrictions of, **2**:28–30

Field goals, measurement of, **3**:124–125

Fifty-five mph
   reasons for speed limit of, **2**:143
   speedometers and, **2**:144–145

Fighting Irish, Notre Dame and, **10**:115–117

Figure skating and dizziness, **5**:33–35

"Filibuster," **11**:110

"Filipino," spelling of, **1**:171

Film
   countdown leader on, **2**:9; **3**:238
   measurement of, in millimeters, **1**:44
   number of exposures in, **1**:153–154
   prints, after theatrical runs, **5**:84–85
   wagon wheels in, movement of, **2**:183
   width of, **1**:222–223
   X- and XXX-ratings, **2**:141–142

Filters, cigarette, brown spots on, **6**:112–113

"Filthy lucre," **11**:117

"Fin [five-dollar bill]," **11**:111

Fingernails
   biting of, **8**:183; **9**:223–228
   growth of, versus toenails, **3**:123
   growth of, after death, **4**:163–164
   lunula on, **2**:218; **3**:241
   yellowing of, and nail polish, **7**:129–130

Fingers, length of, **5**:40–41

"Fink," **11**:73

Fire, crackling sound of, **4**:11

Fire extinguishers, expiration date on, **4**:68

Fire helmets, shape of, **10**:199–203

Food labels
 "FD&C" on label, **4**:163
 lack of manufacturer street
  addresses, **4**:85
Football
 barefoot kickers in, **4**:190–
  191
 college, redshirting in, **7**:46–
  48
 distribution of game balls,
  2:44
 goalposts, tearing down of,
  7:181; **8**:213–217
 measurement of first-down
  yardage, **5**:128–129
 origins of "hut" in, **9**:294–
  295
 Pittsburgh Steelers' helmet
  emblems, **7**:67–68
 shape of, **4**:79–81
 sideline population in, **10**:51–
  53
 two-minute warning and,
  10:150–151
 yardage of kickers in, **3**:124–
  125
"Fore," origins of golf expression, **2**:34
Forewords in books, versus introductions and prefaces,
  1:72–73
Forks, switching hands to use,
  3:198
"Fortnight," **11**:194
Fraternities, Greek names of,
  10:94–98
"Freebooter," **11**:110
Freezer compartments, location
  of, in refrigerators, **2**:230–
  231

Freezers
 ice trays in, **10**:92–93
 lights in, **10**:82–85
French dry cleaning, **3**:164–165
French horns, design of, **5**:110–
  111
"French" bread versus "Italian," **7**:165–166; **8**:261–262
"French" style green beans,
  10:125–126
Frogs
 eye closure when swallowing,
  6:115–116
 warts and, **10**:121–123
Frogs of violins, white dots on,
  4:164–165; **9**:291
Frostbite, penguin feet and,
  1:217–218
Fruitcake, alleged popularity of,
  4:197; **5**:205–209; **6**:253;
  7:226–227; **9**:274–276
"Fry," **11**:179
Fuel gauges in automobiles,
  6:273
"Fullback," **11**:138
Full-service versus self-service,
  pricing of, at gas stations,
  1:203–209
Funeral homes, size of, **8**:152–
  155
Funerals
 burials without shoes, **7**:53–
  54
 depth of graves, **7**:14–15
 head position of, in caskets,
  6:8–9
 orientation of deceased,
  7:106
 perpetual care and, **2**:221–
  222

Gondolas, black color of, **4**:86–87

"Good Friday," origins of term, **8**:108–109; **10**:265

Goofy
identity of, **3**:64–65
marital status of, **7**:102

Goofy, Jr., origins of, **7**:102

Goosebumps, faces and, **2**:8–9

Gorillas, chest pounding of, **8**:53–55

Gowns, and caps, at graduations, **6**:99–102

Grades in school, E, **3**:198; **4**:206–209

Grading of cigarettes, **6**:112–113

Grading of eggs, meaning of, **4**:136–137

Graduations, military academy, **2**:20–21

"Grandfather" clock, origins of term, **4**:178

Grape jellies, color of, **7**:142–143

Grapefruit, sweetness of, canned versus fresh, **1**:199

"Grape-nuts," **11**:171

Grapes, raisins and, **2**:218–219

Gravel and placement on flat roofs, **6**:153–154

Graves, depth of, **7**:14–15

Gravy skin loss, when heated, **6**:58

"Gravy train," **11**:62

Grease, color of, **5**:182

Grecian Formula, process of, **8**:110–111

Greek names of fraternities and sororities, **10**:94–98

Green beans, "French" style, **10**:125–126

Green color of glow-in-the-dark items, **9**:139–141

Green lights, versus red lights, on boats and airplanes, **4**:152–153

"Green with envy," **11**:188

"Green" cards, color of, **7**:61–63

"Greenhorn," **11**:189

Greeting cards, shape of, **6**:70–71

Gretzky, Wayne, hockey uniform of, **2**:18; **10**:279

Grimace, identity of McDonald's, **7**:173

Grocery coupons, cash value of, **5**:7–9

Grocery sacks, names on, **2**:166–167

Grocery stores and check approval, **8**:245–246

Groom, carrying bride over threshold by, **4**:159

Growling of stomach, causes of, **4**:120–121

Guitar strings, dangling of, **8**:11–13; **10**:276

Gulls, sea, in parking lots, **6**:198–199; **10**:254–256

Gum, chewing
water consumption and hardening of, **10**:236–237
wrappers of, **8**:111–112

"Gunny sacks," **11**:195

"Guy," **11**:151–152

"Habit [riding costume]," **11**:123

"Hackles," **11**:6–7

Hail, measurement of, **5**:203; **6**:239–240; **7**:234–235; **8**:236–237

Hair
blue, and older women, **2**:117–118
growth of, after death, **4**:163–164
length of, in older women, **7**:179; **8**:192–197
mole, color of, **8**:167–169
parting, left versus right, **1**:116

Hair color, darkening of, in babies, **10**:209

Hair spray, unscented, smell of, **2**:184

Hairbrushes, length of handles on, **7**:38–39

Hairs in mouth, gagging on, **7**:76–77

Hairy ears in older men, **2**:239; **3**:231–233; **5**:227–228

Half dollars, vending machines and, **3**:54–56

"Halfback," **11**:138

Half-mast, flags at, **10**:36–38

Half-moon versus quarter moon, **7**:72–73

Half-numbers in street addresses, **8**:253

Halibut, coloring of, **3**:95–96

Halloween, Jack-o'-lanterns and, **4**:180–181

Halogen lightbulbs, touching of, **5**:164

Ham
checkerboard pattern atop, **7**:66–67
color of, when cooked, **7**:15–16
Easter and consumption of, **1**:151–152

"Ham [actor]," **11**:170–171

Hamburger buns, bottoms of, **2**:32–34

"Hamburger," origins of term, **4**:125

"Hamfatter," **11**:170–171

Hand dryers in bathrooms, **8**:174–176; **10**:266–267

Hand positions in old photographs, **7**:24–26

Handles versus knobs, on doors, **7**:148–149

Handwriting, teaching of cursive versus printing, **7**:34–37

"Hansom cab," **11**:63

Happy endings, crying and, **1**:79–80

Hard hats
backward positioning of, in ironworkers, **4**:94
exterminators and, **2**:51

Hard-boiled eggs, discoloration of, **3**:34

Hat tricks, in hockey, **2**:165–166

Hats
cowboy, dents on, **5**:6; **6**:274; **7**:249–250
declining popularity, **5**:202; **6**:227–231; **7**:233; **7**:249
holes in sides of, **5**:126
numbering system for sizes, **4**:110

Haystacks, shape of, **6**:47–48; **8**:265–266

Holes (*continued*)
in ice cream sandwiches, **8**:128
in needles and syringes, **10**:57–59
in pasta, **4**:28
in saltines, **8**:129
in thimbles, **10**:63–64
in wing-tip shoes, **8**:44
on bottom of soda bottles, **6**:187–188
recycling of, in loose-leaf paper, **7**:105–106
refilling of dirt, **7**:48–49
"Holland," versus "Netherlands," **2**:65–66
Home plate, shape of, in baseball, **5**:131
"Honcho," **11**:39
Honey, spoilage of, **4**:177–178
Honey roasted peanuts, banning of, on airlines, **4**:13–14
Honking in geese during migration, **7**:108
"Honky," **11**:77
"Hoodwink," **11**:121
"Hoosiers," **11**:148–149
"Horsefeathers," **11**:40
Horses
measurement of heights of, **5**:60–61
posture in open fields, **3**:104; **5**:241
shoes, **3**:156
sleeping posture, **2**:212
vomiting, **6**:111–112; **7**:248
Hospital gowns, back ties on, **5**:132–134

Hospitals and guidelines for medical conditions, **4**:75–76
Hot dog buns
number of, in package, **2**:232–235
slicing of, **5**:161
Hot dogs, skins of, **5**:54
Hot water, noise in pipes of, **2**:199–200
Hotels
amenities, spread of, **6**:118–121
number of towels in rooms, **4**:56–57
plastic circles on walls of, **3**:117
toilet paper folding in bathrooms of, **3**:4
"Hotsy totsy," **11**:40
Houses, settling in, **6**:32–34
"Hue and cry," **11**:112
"Humble pie," **11**:169
Humidity, relative, during rain, **1**:225–226
Humming, power lines and, **10**:259
Hurricane, trees and, **3**:68–69
"Hurricanes" as University of Miami nickname, **8**:171–172
"Hut," origins of football term, **6**:40; **9**:294–295
Hydrants, fire, freezing water in, **10**:11
Hypnotists, stage, techniques of, **1**:180–191

"I [capitalization of]," **11**:55
"I could care less," **11**:78

"I" before "e," in spelling, **6**:219; **7**:209; **8**:240–245

Ice

fizziness of soda, **9**:24–25

formation on top of lakes and ponds, **5**:82–83

holes and dimples in, **9**:147–148

in urinals, **10**:232–234

Ice cream

black specks in, **8**:132–133

cost of cones versus cups, **1**:133–135

pistachio, color of, **7**:12–13

thirstiness, **5**:202; **6**:236–237

Ice cream and soda, fizziness of, **9**:27

Ice cream sandwiches, holes in, **8**:128

Ice cubes

cloudy versus clear, **3**:106–107; **5**:242

shape of, in home freezers, **5**:103–104

Ice rinks, temperature of resurfacing water in, **10**:196–198

Ice skating, awful music in, **1**:102–105

Ice trays in freezers, location of, **10**:92–93

Icy roads, use of sand and salt on, **2**:12–13

Ignitions

automobile, and headlamp shutoff, **7**:92–93

key release button on, **5**:169

Imperial gallon, versus American gallon, **6**:16–17

"In like Flynn," **11**:157

"In the nick of time," **11**:158

Index fingers and "Tsk-Tsk," stroking of, **4**:198; **5**:209–210

"Indian corn," **11**:146

"Indian pudding," **11**:146

"Indian summer," **11**:146

Indianapolis 500, milk consumption by victors in, **8**:130–131

"Inflammable," versus "flammable," **2**:207–208

Ink

color of, in ditto masters, **6**:133–134

newspaper, and recycling, **7**:139–140

Insects

attraction to ultraviolet, **8**:158–159

aversion to yellow, **8**:158–159

flight patterns of, **7**:163–164

in flour and fruit, **4**:89–90

See also specific types

Insufficient postage, USPS procedures for, **4**:149–151

Interstate highways, numbering system of, **4**:66–67

Introductions in books, versus forewords and prefaces, **1**:72–73

Irish names, "O'" in, **8**:135–136

Irons, permanent press settings on, **3**:186–187

Ironworkers, backwards hard hat wearing of, **4**:94

Irregular sheets, proliferation of, **1**:145–147

IRS and due date of taxes, **5**:26–29

IRS tax forms
 disposal of, **8**:143–144
 numbering scheme of, **4**:9–10
"Italian" bread, versus "French," **7**:165–166; **8**:261–262
Itching, reasons for, **1**:172–173
Ivory soap, purity of, **2**:46–47

"J" Street, Washington, D.C., and, **2**:71
"Jack [playing card]," **11**:135
Jack Daniel's and "Old No. 7," **8**:144–145
"Jack," "John" versus, **2**:43
Jack-o'-lanterns, Halloween and, **4**:180–181
Jams, contents of, **6**:140–141
Japanese
 baseball uniforms, **10**:207–208
 boxes, yellow color of, **7**:130–131
 flags, red beams and, **10**:151–155
Jars, food, refrigeration of opened, **6**:171–172
"Jaywalking," **11**:22–23
Jeans
 blue, orange thread and, **9**:74
 Levi's, colored tabs on, **6**:59–61
 origin of "501" name, **6**:61
 sand in pockets of new, **7**:152
"Jeans [pants]," **11**:124
"Jeep," **11**:61
Jellies, contents of, **6**:140–141
Jellies, grape, color of, **7**:142–143

Jello-O, fruit in, **3**:149–150
Jeopardy, difficulty of Daily Doubles in, **1**:33–35
"Jerkers," **11**:176
Jet lag, birds and, **3**:33–34
"Jetsam," versus "flotsam," **2**:60–61
"Jig is up," **11**:7
Jigsaw puzzles, fitting pieces of, **9**:3–4
Jimmies, origins of, **10**:165–168
"Jink," **11**:41
"John," versus "Jack," **2**:43
Johnson, Andrew, and 1864 election, **8**:85–87
"Joshing," **11**:158–159
Judges and black robes, **6**:190–192
Judo belts, colors of, **9**:119–123
"Juggernaut," **11**:63–64
Juicy Fruit gum, flavors in, **1**:71; **3**:242

"K rations," **11**:54–55
"K" as strikeout in baseball scoring, **5**:52–53
Kangaroos, pouch cleaning of, **4**:144–145
Karate belts, colors of, **9**:119–123
"Keeping up with the Joneses," **11**:159–160
Ken, hair of, versus Barbie's, **7**:4–5; **8**:259–260
Ketchup, Heinz, labels of, **8**:150–151
Ketchup bottles
 narrow necks of, **2**:44–45
 neck bands on, **5**:242
 restaurants mating of, **3**:200

"Ketchup," **11**:177

"Kettle of fish," **11**:178

Keys

  automobile, door and ignition, **3**:141–142

  piano, number of, **10**:7–9

  teeth direction, **8**:59–60

  to cities, **3**:99

"Kidnapping," **11**:113–114

Kids versus adult goats, **7**:64–65

Kilts, Scotsmen and, **7**:109–110

Kissing

  eye closure during, **7**:179; **8**:186–191

  leg kicking by females, **6**:218; **7**:196–197; **9**:278; **9**:299–300

"Kit cat club," **11**:38–39

"Kit," "caboodle" and, **2**:15–17

"Kittycorner," **11**:197

Kiwifruit in gelatin, **3**:149–150

Kneading and bread, **3**:144–145

Knee-jerk reflex in humans, **8**:255–256

Knives, dinner, rounded edges of, **1**:231–232

Knives, serrated, lack of, in place settings, **4**:109–110

Knobs versus handles, on doors, **7**:148–149

"Knock on wood," **11**:4–5

"Knuckle down," **11**:9–10

"Knuckle under," **11**:9

Knuckles, wrinkles on, **5**:182–183

Kodak, origins of name, **5**:169–170; **9**:290

Kool-Aid and metal containers, **8**:51

Kraft American cheese, milk in, **1**:247–249

"L.S.," meaning of, in contracts, **1**:165

Label warnings, mattress, **2**:1–2

Labels on underwear, location of, **4**:4–5

Labels, food, lack of manufacturer street addresses on, **4**:85

"Ladybug," **11**:23

Ladybugs, spots on, **7**:39–40

Lakes

  effect of moon on, **5**:138–139

  fish returning to dried, **3**:15–16; **10**:256

  ice formations on, **5**:82–83

  versus ponds, differences between, **5**:29–30; **7**:241

  versus ponds, water level of, **9**:85–86

  wind variations, **4**:156–157

"Lame duck," **11**:24–25

Lane reflectors, fastening of, **5**:98–99

Large-type books, size of, **5**:135

Laryngitis, dogs, barking, and, **2**:53–54

Lasagna, crimped edges of, **5**:61

"Last ditch," **11**:10

"Last straw," **11**:8–9

Laughing hyenas, laughter in, **8**:1–2

Lawn ornaments, plastic deer as, **8**:185

Lawns, reasons for, **2**:47–50

"Lawyer," **11**:103–104

"Lb. [pound]," **11**:56

Leader, film, **2**:9
"Leap year," **11**:193
Leather, high cost of, **8**:21–23
Ledges in buildings, purpose of, **8**:18–20
Left hands, placement of wristwatches on, **4**:134–135; **6**:271
"Left wing," **11**:116
Left-handed string players, in orchestras, **9**:108–109; **10**:276–277
Leg kicking by women while kissing, **6**:218; **7**:196–197; **9**:278; **9**:299–300
Legal-size paper, origins of, **3**:197
"Legitimate" theater, origins of term, **10**:5–7
"Let the cat out of the bag," **11**:25
Letters
    business, format of, **7**:180; **8**:201–204
    compensation for, between countries, **4**:5–6
Letters in alphabet soup, distribution of, **3**:118–119
Levi's jeans
    colored tabs, **6**:59–61
    origin of "501" name, **6**:61
Liberal arts, origins of, **5**:70–73
Lice, head, kids and, **10**:225–227
License plates and prisoners, **8**:137–139; **10**:268–269
License plates on trucks, absence of, **3**:98; **10**:270–271
"Licking his chops," **11**:47
Licorice, ridges on, **9**:188–189

Life Savers, wintergreen, sparkling of, when bitten, **1**:157–158
Lightbulbs
    air in, **6**:199–200
    fluorescent, stroking of, **3**:131
    halogen, **5**:164
    high cost of 25-watt variety, **5**:91
    in traffic signals, **3**:31–32
    loosening of, **3**:93–94
    noise when shaking, **5**:167
    plinking by fluorescent, **5**:47
    three-way, burnout, **2**:104
    three-way, functioning of, **2**:105
    wattage sizes of, **1**:255–256
Light switches, height and location of, **4**:183–184
Lighters, disposable, fluid chambers of, **6**:92–93
Lightning, heat, **2**:185
Lights in freezers, **10**:82–85
"Limelight," **11**:33
Lincoln, Abraham, and Andrew Johnson, **8**:85–87
Linens, unpleasant smell in, **8**:57–58
Lions, animal trainers and, **7**:9–11
Lips
    black, on dogs, **3**:38–39
    upper, grooves on, **6**:42–43
Liquids as treatment for colds, **4**:131–132
Liquor, proof and, **2**:177
Liquor distilleries during Prohibition, **9**:54–56

Manhole covers, round shape of, **3**:191

Marching, stepping off on left foot when, **5**:172–173; **9**:293–294

Marching bands, formations of, **8**:107–108

Margarine, versus butter, in restaurants, **1**:32–33

Margarine sticks, length of, **5**:42

Marmalades, contents of, **6**:140–141

Marshals' badges, shape of, **5**:73–74

Marshmallows, invention of, **8**:99–100

Martial arts, sniffing and, **10**:256–258

Martinizing, One Hour, **3**:28–29

Mascara, mouth opening during application of, **1**:257–260

Matchbooks, location of staples on, **6**:173–174

Matches, color of paper, **6**:174–175

Math, school requirement of, **8**:184; **9**:254–261

Mattress tags, warning labels on, **2**:1–2

Mattresses, floral graphics on, **9**:1–2

Maximum occupancy in public rooms, **10**:158–160

Mayonnaise, Best Foods versus Hellmann's, **1**:211–214

Mayors, keys to cities and, **3**:99

McDonald's
Grimace, identity of, **7**:173
"over 95 billion served" signs, **7**:171

sandwich wrapping techniques, **7**:172

straw size, **7**:171–172

Measurements
acre, **2**:89
meter, **2**:200–202

Measuring spoons, inaccuracy of, **1**:106–107

Meat
children's doneness preferences, **5**:230–231; **6**:252–253; **9**:273–274
national branding, **1**:227–231; **9**:287
red color of, **8**:160–161

Meat loaf, taste in institutions, **5**:203; **6**:243; **7**:235–236

Medals, location of on military uniforms, **2**:223–224

Medical conditions, in hospitals, guidelines for, **4**:75–76

Medicine bottles, cotton in, **3**:89–90

Memorial Day, Civil War and, **3**:168–169

Men
dancing ability of, **6**:218; **7**:199–202; **8**:239–240
feelings of coldness, **6**:218; **7**:198–199
remote controls and, **6**:217; **7**:193–196

Menstruation, synchronization of, in women, **4**:100–102

Menthol, coolness of, **5**:192

Meter, origins of, **2**:200–202

Miami, University of
football helmets, **8**:171–172
"Hurricanes" nickname, **8**:171–172

Movie actors and speed of speech, **5**:203; **6**:241–243

Movie theaters
bells in, **1**:88–89
in-house popcorn popping, **1**:45–50

Movies, Roman numerals in copyright dates in, **1**:214–216

"Mrs.," **11**:57

MSG, Chinese restaurants and, **2**:168–171

"Mugwump," **11**:119

Muppets, left-handedness of, **7**:111–113

Murder scenes, chalk outlines at, **3**:11–12

Musketeers, Three, lack of muskets of, **7**:29–30

Mustaches, policemen and, **6**:219; **7**:218–220; **8**:246–247; **9**:278

"Muumuu," **11**:125

"Mystery 7," in *$25,000 Pyramid*, **1**:192

Nabisco Saltine packages, red tear strip on, **1**:147–149

Nabisco Shredded Wheat box, Niagara Falls on, **5**:100–101

Nail polish and fingernail yellowing, **7**:129–130

"Namby pamby," **11**:79

National Geographics, saving issues of, **3**:199; **5**:229–230; **7**:224

Navy and Army, Captain rank in, **3**:48–50

Necklaces and clasp migration, **7**:180; **8**:197–201; **9**:279–281

Neckties
direction of stripes on, **6**:86–87
origins of, **4**:127; **8**:264–265
taper of, **6**:84–85

Nectarines, canned, lack of, **4**:59–60; **9**:287–288

Needles, holes in, of syringes, **10**:57–59

Neptune's moon, Triton, orbit pattern of, **4**:117–118

Nerdiness and eyeglasses, **7**:180

"Netherlands," versus "Holland," **2**:65–66

New York City and steam in streets, **5**:16–17

"New York" steaks, origins of, **7**:155–156; **8**:252

New Zealand, versus "Old Zealand," **4**:21–22

Newspapers
ink and recycling of, **7**:139–140
ink smudges on, **2**:209–212
jumps in, **5**:116–117
tearing of, **2**:64
window cleaning and, **10**:33–36
yellowing of, **8**:51–52

Niagara Falls on Nabisco Shredded Wheat box, **5**:100–101

"Nick of time," **11**:158

Nickels, smooth edges of, **1**:40–41

"Nickname," **11**:163

Nightclubs, lateness of bands in, **8**:184; **9**:248–254

"Nine-day wonder," **11**:98

Nine-volt batteries, shape of, **6**:104; **7**:242–243

Nipples, purpose of, in men, **4**:126; **6**:275

"No bones about it," **11**:49

"No Outlet" signs, versus "Dead End" signs, **4**:93

Noise, traffic, U.S. versus foreign countries, **4**:198

North Carolina, University of, and Tar Heels, **8**:76–77

North Pole
directions at, **10**:243
telling time at, **10**:241–243

Nose rings and bulls, **10**:147–148

Noses
clogged nostrils and, **3**:20–21
runny, in cold weather, **10**:146–147
runny, kids versus adults, **9**:89–90
wet, in dogs, **4**:70–73

Nostrils, clogged, **3**:20–21

Notches on bottom of shampoo bottles, **10**:29–30

Notre Dame fighting Irish, **10**:115–117

NPR radio stations, low frequency numbers of, **10**:181–183

Numbers, Arabic, origins of, **3**:16–17

Nutrition labels, statement of fats on, **6**:142–143

Nuts
Brazil, in assortments, **7**:145–147
Macadamia shells, **8**:262
peanuts in plain M&M's, **7**:239
peanuts, and growth in pairs, **7**:34

"O'" in Irish names, **8**:135–136

Oakland A's, elephant on uniforms of, **6**:14–15

Oboes, use of as pitch providers, in orchestras, **4**:26–27

Occupancy, maximum, in public rooms, **10**:158–160

Oceans
boundaries between, **10**:74–76
color of, **2**:213
salt in, **5**:149–150
versus seas, **5**:30–32

Octopus throwing, Detroit Red Wings and, **9**:183–186

"Off the schneider," **11**:136

Oh Henry!, origins of name of, **8**:83–84

Oil
automotive, after oil change, **5**:184–185; **7**:240–241
automotive, grades of, **3**:182–183

"Okay," thumbs-up gesture as, **1**:209–210

Oktoberfest, September celebration of, **9**:156–157

Old men
hairy ears and, **2**:239; **3**:231–233; **5**:227–228

Paperback books, staining of, **2**:93–94

Papers, yellowing of, **8**:51–52

"Par [golf course]," **11**:139–140

"Par Avion" on air mail postage, **8**:39

"Pardon my French," **11**:150

Parking lots, sea gulls at, **10**:254–256

Parking meters, yellow "violation" flags and, **4**:42–43

"Parkway," **11**:65

Parkways, parking on, versus driveways, **4**:123

Parrots and head bobbing, **8**:23–24

Parting of hair, left versus right, **1**:116

Partly cloudy, versus partly sunny, **1**:21–22

Partly sunny, versus partly cloudy, **1**:21–22

"Pass the buck," **11**:107

Pasta
boxes, numbers on, **4**:107
foaming when boiling, **7**:78
holes in, **4**:28

Pay phones, collection of money from, **1**:107–108

Pay toilets, disappearance of, **2**:25–26

PBX systems, **3**:75–76

"Pea jacket," **11**:124

Peaches
canned, and pear juice, **8**:46–47
fuzziness of, **4**:58–59

Peanut butter, stickiness of, **10**:204–207

Peanuts
allergies to, **7**:239
growth in pairs, **7**:34
honey roasted, and airlines, **4**:13–14
origins of comics name, **10**:191–193

Pear juice in canned peaches, **8**:46–47

Pears and apples, discoloration of, **4**:171

Pebbles, spitting by fish of, **9**:174–175

"Peeping Tom," **11**:162

Pencils
architectural and art, grades of, **7**:173
carpenter's, shape of, **7**:27; **9**:290–291
color, **3**:108
numbering, **3**:109

Penguins
frostbite on feet, **1**:217–218
knees, **5**:160

Penicillin and diet, **8**:95–96

Penmanship of doctors, bad, **5**:201; **6**:221–225; **7**:232–233; **8**:235

Pennies
lettering on, **7**:5
smooth edges of, **1**:40–41
vending machines and, **3**:54–56

Pennsylvania Dept. of Agriculture, registration, baked goods, **2**:121–122

Penny loafers, origins of, **8**:43–44

Pine trees, construction sites and, **2**:147–148

Pineapple in gelatin, **3**:149–150

Pinholes, on bottle caps, **2**:223

Pink as color for baby girls, **1**:29

"Pink lady," **11**:190–191

Pink stripes on magazine labels, **8**:96–97

"Pinkie," **11**:190–191

Pins in men's dress shirts, **4**:29–30

"Pipe down," **11**:13

Pipes, kitchen, shape of, **4**:82–83

Pirates
  earrings on, **9**:43–45; **10**:272–273
  walking the plank, **9**:37–42; **10**:273–274

Pistachio ice cream, color of, **7**:12–13

Pistachios, red color of, **1**:26–28

Pita bread, pockets in, **6**:98

Pitcher's mound
  location of, **5**:181
  rebuilding of, **9**:195–198

Pittsburgh Steelers, emblems on helmets of, **7**:67–68

Planets, twinkling of, at night, **4**:50–51

Plastic bottles, beer and, **7**:161–162

Plastic cups, shape of, **9**:289

Plastic deer ornaments on lawns, **9**:262–264

Plastics, recyclable, numbers on, **6**:155–156

Plates
  repositioning of, **8**:184; **9**:238–242

round shape of dinner, **8**:162–164

Plots, farm, circular shape of, **7**:118–119

Plug prongs
  holes at end of, **5**:94–95
  three prongs versus two prongs, **5**:191

Plum pudding, plums in, **5**:49

Plumbing
  kitchen, shape of, **4**:82–83
  sound of running water, **3**:239–240

Pockets in pita bread, **6**:98

Poison ivy, grazing animals and, **3**:86–87

Polaroid prints, flapping of, **7**:175–176

Poles
  directions at, **10**:243
  telling time at North and South, **10**:241–243

Pole-vaulting
  preparation for different heights, **9**:97–101
  women and, **9**:102–107

Police
  and crowd estimates, **1**:250–253
  flashlight grips, **10**:30–32
  radar and speed measurement, **8**:88–91

Police car beacons, colors on, **7**:135–137

Police dogs, urination and defecation of, **3**:67–68

Policemen and mustaches, **6**:219; **7**:218–220; **8**:246–247; **9**:278

Ponds
    effect of moons on, **5**:138–139
    fish returning to dried, **3**:15–16; **10**:256
    ice formations on, **5**:82–83
    versus lakes, **5**:29–30; **7**:241
    versus lakes, level of, **9**:85–86
Poodles, wild, **6**:207–209
Pool balls, dots on, **10**:237–240
"Pop goes the weasel," **11**:14
Popcorn
    "gourmet" versus regular, **1**:176
    popping in-house, in movie theaters, **1**:45–50
    versus other corns, **3**:142–143
Popes
    name change of, **10**:17–18
    white skullcap of, **10**:80–81
    white vestments of, **10**:79–80
Popping noise of wood, in fires, **4**:10
Pork and beans, pork in, **2**:19
"Port," **11**:65–66
"Porthole," **11**:65–66
Postage and ripped stamps, **8**:62
Postage stamps
    leftover perforations of, **4**:179
    taste of, **2**:182
Postal Service, U.S., undeliverable mail and, **5**:13–14
Pot pies, vent holes in, **6**:28
Potato chips
    bags, impossibility of opening and closing, **9**:117–118
    curvy shape, **9**:115–116
    green tinges on, **5**:136–137; **6**:275

price of, versus tortilla chips, **5**:137–138
Potato skins, restaurants and, **3**:12–13
Potatoes, baked, and steak houses, **6**:127–129
Potholes, causes of, **2**:27
"Potter's field," **11**:198
Power lines
    humming of, **9**:165–168; **10**:259
    orange balls on, **4**:18–19
Prefaces in books, versus introductions and forewords, **1**:72–73
Pregnancy, permanents and, **3**:170–171
Pregnant women, food cravings of, **10**:183–185
Preserves, contents of, **6**:140–141
Press conferences, microphones in, **2**:11–12
"Pretty kettle of fish," **11**:178
"Pretty picnic," **11**:178
Pretzels, shape of, **6**:91–92
Priests, black vestments and, **10**:77–79
Priority mail, first class versus, **3**:166–167
Prisoners and license plate manufacturing, **8**:137–139
Prohibition, liquor production of distilleries during, **9**:54–56
Pronunciation, dictionaries and, **10**:169–179
"P's and Q's," **11**:88–89
Pubic hair
    curliness of, **5**:177–178

purpose of, **2**:146; **3**:242–243; **6**:275–276

Public buildings, temperatures in, **8**:184

Public radio, low frequency numbers of, **10**:181–183

Pudding, film on, **6**:57

Punts, measurement of, in football, **3**:124–125

Purple
Christmas tree lights, **6**:185–186; **9**:293; **10**:280
paganism, **9**:292–93
royalty and, **6**:45–46

"Put up your dukes," **11**:137

Putting, veering of ball toward ocean when, **6**:107–108

"Q.T.," **11**:59

"Qantas," spelling of, **8**:134–135

Q-Tips, origins of name, **6**:210–211

"Quack [doctor]," **11**:45

"Quarterback," **11**:138

Quarterbacks and exclamation, "hut," **6**:210

Quarter-moons versus half moons, **7**:72–73

Quarts and gallons, American versus British, **4**:114–115

Queen-size sheets, size of, **3**:87–88

Rabbit tests, death of rabbits in, **7**:69–71

Rabbits and nose wiggling, **5**:173–174

Racewalking, judging of, **9**:20–23

Racquetballs, color of, **4**:8–9

Radar and police speed detection, **8**:88–91

Radiators and placement below windows, **9**:128–130

Radio
beeps before network news, **1**:166–167
FM, odd frequency numbers of, **10**:59–60
lack of song identification, **7**:51–57
public, low frequency numbers, **10**:181–183

Radio Shack and lack of cash registers, **5**:165–166

Radios
battery drainage, **10**:259–260
lingering sound of recently unplugged, **4**:47

Railroad crossings and "EXEMPT" signs, **5**:118–119

Railroads, width of standard gauges of, **3**:157–159

Rain
butterflies and, **4**:63–64
fish biting in, **10**:131–138
measurement container for, **4**:161–163
smell of impending, **6**:170–171; **7**:241

Raincoats, dry-cleaning of, **2**:216–217

"Raining cats and dogs," **11**:26

"Raise hackles," **11**:6–7

Raisins
cereal boxes and, **2**:123
seeded grapes and, **2**:218–219

Shampoo bottles, notches on
bottom of, **10**:29–30
Shampoo labels, "FD&C" on
label of, **4**:163
Shampoos
colored, white suds and,
4:132–133
lathering of, **5**:44–45
number of applications, **1**:90–
93
Shaving of armpits, women and,
2:239; **3**:226–229; **6**:249
Sheets
irregular, proliferation of,
1:145–147
queen-size, size of, **3**:87–88
Sheriffs' badges, shape of, **5**:73–
74
Shirts
buttons on men's versus
women's, **2**:237–238;
3:207–209; **5**:226; **7**:223
men's, pins in, **4**:29–30
single-needle stitching in,
6:51
starch on, **3**:118
Shoe laces
length in athletic shoes,
8:41–42
untied, in shoe stores, **8**:40
Shoe sizes, differences between, **1**:65–70
Shoes
lace length in shoe stores,
8:41–42
layers on, **5**:59
of deceased, at funerals,
7:153–154
penny loafers, **8**:43–44
single, on side of road,

2:236–237; **3**:201–207;
4:232–233; **5**:225–226;
6:245–248; **7**:221–222;
8:228–230; **9**:271–272
tied to autos, at weddings,
1:235–238
uncomfortable, and women,
1:62–64
untied laces in stores, **8**:40
wing-tip, holes in, **8**:44
"Shoofly pie," **11**:179
Shopping, female proclivity toward, **7**:180; **8**:205–209
Shopping malls, doors at entrance of, **6**:180–181
"Short shrift," **11**:15
Shoulder straps and seat belts
in airplanes, **8**:141–142
Shredded Wheat packages,
Niagara Falls on, **5**:100–
101
"Shrift," **11**:15
Shrimp, baby, peeling and
cleaning of, **5**:127
"Shrive," **11**:15
"Siamese twins," **11**:151
"Sic," as dog command, **5**:51
Side vents in automobile win-
dows, **6**:13–14
"Sideburns," **11**:126–127
Sidewalks
cracks on, **3**:176–178
glitter on, **7**:160; **10**:61–62
Silica gel packs in electronics
boxes, **6**:201
Silos, round shape of, **3**:73–74;
5:245; **10**:260–261
Silver fillings, rusting of, **10**:41–
42
Silverstone, versus Teflon, **2**:3

Teeth direction of keys, **8**:59–60

"Teetotaler," **11**:42–43

Teflon, stickiness of, **2**:3

Telegrams
exclamation marks and, **3**:76–77
periods and, **3**:77–78

Telephone cords, twisting of, **3**:45

Telephone rings, mechanics of, **4**:189–190

Telephones
911 as emergency number, **5**:145–146
area code numbers, **5**:68–69; **9**:287
dialing 9 to get outside line, **3**:75–76
fast versus slow busy signals, **4**:182
holes in mouthpiece, **3**:14–15
pay, clicking noise in, **4**:97–98
"Q" and "Z," absence from buttons, **5**:66–67
rings, mechanics of, **4**:189–190
third-party conversations, **9**:152–155
three-tone signals, **4**:129–130
touch tone keypad for, **2**:14–15
unlisted phone numbers, **9**:45–47
windowless central offices, **9**:176–181

Telescopes, inverted images of, **3**:50–51

Television
advertising sales for overrun live programming and, **1**:50–52
cable and channel allocation, **9**:75–76; **10**:267
channel 1 on, lack of, **7**:242; **10**:267–268
"snow" on, **9**:199–200
sound of, in bars, **10**:12–14
volume levels of, **5**:57–58

Television commercials, loudness of, **3**:81–83

Television sets
diagonal measurement of, **4**:37; **5**:246
measurement of, Canadian versus U.S., **6**:23

Temperature
babies and tolerance for high, **4**:103–104
cold water, kitchen versus bathroom, **4**:151–152
human comfort and, **2**:178–179
in public buildings, **8**:184; **9**:243–247
perception of air versus water, **4**:184

Ten percent of brain, alleged use of, **4**:198; **5**:210–211; **6**:254–256; **8**:232

"Ten-foot pole," **11**:99

Tennis, scoring in, **2**:3–5

Tennis balls
fuzziness of, **3**:35–36
high-altitude, **8**:80

"Tenterhooks," **11**:10–11

Tequila, worms in bottles of, **7**:88–89; **9**:289

Virgin acrylic, **7**:97–98

Virgin olive oil, **3**:174–175

Vision, 20–20, **3**:143

Vitamins, measurement of, in foods, **6**:148–150

Voices

  causes of high and low, **2**:70

  elderly versus younger, **6**:24–25

  perception of, own versus others', **1**:95–96

Volkswagen Beetles, elimination of, **2**:192–194

Vomiting and horses, **6**:111–112

"Waffling," **11**:183

Wagon wheels in film, movement of, **2**:183

Waiters' tips and credit cards, **7**:133–135

Walking the plank, pirates and, **9**:37–42; **10**:273–274

Walking, race, judging of, **9**:20–23

Wall Street Journal, lack of photographs in, **4**:41–42

Warmth and its effect on pain, **3**:134–135

Warning labels, mattress tag, **2**:1–2

Warts, frogs and toads and, **10**:121–123

Washing machine agitators, movement of, **4**:56

Washing machines, top- versus bottom-loading, and detergent, **1**:159–165

Washington, D.C., "J" Street in, **2**:71

Watch, versus clock, distinctions between, **4**:77–78

"Watch," origins of term, **4**:77

Watches and placement on left hand, **4**:134–135; **6**:271

Water

  bottled, expiration dates on, **9**:77–78

  chemical manufacture of, **5**:107–108

  clouds in tap water, **9**:126–127

  color of, **2**:213

Water faucets

  bathroom versus kitchen, **5**:244

  location of, hot versus cold, **4**:191–192

Water temperature

  effect on stain, **6**:77–78

  versus air temperature, perception of, **4**:184

Water towers

  height of, **5**:91–93

  winter and, **6**:38–40

Water, boiling during home births, **6**:114–115; **7**:247–248; **8**:254

Water, cold, kitchen versus bathroom, **4**:151–152

Watermelon seeds, white versus black, **5**:94

Wax, whereabouts in dripless candles, **4**:182–183

"Wear his heart on his sleeve," **11**:128

"Weasel words," **11**:90–91

Weather

  clear days following storms, **6**:125

fainting, **10**:219–222

feelings of coldness, **6**:218; **7**:198–199; **8**:238

group restroom visits, **6**:217; **7**:183–192; **8**:237–238; **9**:277–278

hair length of aging, **8**:192–197

leg kicking when kissing, **6**:218; **7**:196–197; **9**:278; 299–300

remote control usage, **6**:217; **7**:193–196

spitting, **8**:226–227

uncomfortable shoes, **1**:62–64

Wood, popping noise of, in fires, **4**:10

Woodpeckers, headaches and, **4**:44–45

Wool

shrinkage of, **6**:166–167

smell of, when wet, **4**:158–159

Worms

appearance on sidewalk after rain, **4**:109

as fish food, **3**:110–112

birds and, **10**:65–72

in tequila bottles, **7**:88–89

larvae and, **6**:269–270

survival during winter, **4**:108

Wrapping

Burger King sandwiches, **8**:112–113

chewing gum, **8**:111–112

gift box chocolate, **8**:122–123

Wrinkles on extremities, **2**:112

Wrists as perfume target, **6**:90

Wristwatches, placement on left hand, **4**:134–135; **6**:271

X

as symbol for kiss, **1**:128–129

as symbol in algebra, **9**:131–132

"X ray," **11**:49–50

"Xmas," origins of term, **2**:75

X-rated movies, versus XXX-rated movies, **2**:141–142

"XXX [liquor]," **11**:58

Yawning, contagiousness of, **2**:238; **3**:213–217; **8**:231–232

Yeast in bread, effect of, **3**:144–145

Yellow, aversion of insects to, **8**:158–159

Yellow Freight Systems, orange trucks of, **6**:64–65

Yellow lights, timing of, in traffic lights, **1**:109–112

Yellow Pages, advertisements in, **3**:60–63

Yellowing of fingernails, nail polish and, **7**:129–130

YKK on zippers, **5**:180

Yogurt

fruit on bottom, **7**:83

liquid atop, **7**:82

Zebras and riding by humans, **8**:139–141

ZIP code, addresses on envelopes and, **3**:44

"Zipper," **11**:129

Zippers, YKK on, **5**:180

Zodiac, different dates for signs in, **4**:27–28

Zoo animals, rocking in, **10**:279

Help!

We hate to end the book on a downbeat note, but we have to admit one dread fact: Imponderability is not yet smitten. Let's stamp it out.

"How?" you ask. Send us letters with your Imponderables, answers to Frustables, gushes of praise, and even your condemnations and corrections.

Join your inspired comrades and become a part of the wonderful world of *Imponderables*. If you are the first person to submit an Imponderable we use in the next volume, we'll send you a complimentary copy, along with an acknowledgment in the book.

Although we accept "snail mail," we strongly encourage you to e-mail us if possible. Because of the volume of mail, we can't always provide a personal response to every letter, but we'll try—a self-addressed stamped envelope doesn't hurt. We're much better with answering e-mail, although we fall far behind sometimes when work beckons.

Come visit us online at the Imponderables website, where

you can pose Imponderables, read our blog, and find out what's happening at Imponderables Central. Send your correspondence, along with your name, address, and (optional) phone number to:

feldman@imponderables.com
www.imponderables.com

Imponderables
P.O. Box 116
Planetarium Station
New York, NY 10024-0116

DO ELEPHANTS JUMP?
ISBN 0-06-053913-5 (hardcover)

Find the answers to perplexing questions and solutions to everyday mysteries in David Feldman's newest book. Includes a cumulative index of all ten Imponderables® books and charming illustrations by longtime collaborator Kassie Schwan.

**HOW DOES ASPIRIN
FIND A HEADACHE?**
ISBN 0-06-074094-9 (paperback)

**ARE LOBSTERS
AMBIDEXTROUS?**
ISBN 0-06-076295-0 (paperback)

COMING SUMMER 2005 FROM DAVID FELDMAN!